D1001891

HISTORY OF
CLASSICAL SCHOLARSHIP
FROM 1300 TO 1850

IN MEMORIAM
UXORIS CARISSIMAE

HISTORY OF
CLASSICAL
SCHOLARSHIP

FROM 1300 TO 1850

RUDOLF PFEIFFER

The King's Library

CLARENDON PRESS · OXFORD

VNYS AZ 201 .P43
1999 c.1

shen031

OXFORD
UNIVERSITY PRESS

Great Clarendon Street, Oxford OX2 6DP

Oxford University Press is a department of the University of Oxford.
It furthers the University's objective of excellence in research, scholarship,
and education by publishing worldwide in

Oxford New York

Athens Auckland Bangkok Bogotá Buenos Aires Calcutta
Cape Town Chennai Dar es Salaam Delhi Florence Hong Kong Istanbul
Karachi Kuala Lumpur Madrid Melbourne Mexico City Mumbai
Nairobi Paris São Paulo Singapore Taipei Tokyo Toronto Warsaw
with associated companies in Berlin Ibadan

Oxford is a registered trade mark of Oxford University Press
in the UK and in certain other countries

Published in the United States
by Oxford University Press Inc., New York

© Oxford University Press 1976

The moral rights of the author have been asserted
Database right Oxford University Press (maker)

Special edition for Sandpiper Books Ltd., 1999

All rights reserved. No part of this publication may be reproduced,
stored in a retrieval system, or transmitted, in any form or by any means,
without the prior permission in writing of Oxford University Press,
or as expressly permitted by law, or under terms agreed with the appropriate
reprographics rights organisation. Enquiries concerning reproduction
outside the scope of the above should be sent to the Rights Department,
Oxford University Press, at the address above

You must not circulate this book in any other binding or cover
and you must impose this same condition on any acquirer

British Library Cataloguing in Publication Data
Data available
ISBN 0-19-814364-8

1 3 5 7 9 10 8 6 4 2

Printed in Great Britain
on acid-free paper by
Bookcraft (Bath) Ltd,
Midsomer Norton

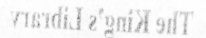

PREFACE

In the preface to the first volume of this *History* I said a few words about the particular problem of reflecting upon the past of classical scholarship and upon the scholars of bygone days. The present volume which is concerned with the period from the thirteenth to the nineteenth century, coming nearly to our own day, may have a stronger appeal to many readers than its predecessor.

In that preface I also acknowledged the help I was fortunate to find in writing this *History*. In the first place I have to mention again Mr. E. Arnold, now in the Manuscript Department of the Bavarian State Library; when my forces began to decline in my eighties, he redoubled his efforts and by doing so has made it possible for this volume to be published. The annual grants voted to me by the Bayerische Akademie der Wissenschaften and by the British Academy have enabled me to meet the expense of this invaluable assistance through so many years.

Of my many Oxford and Munich friends, who offered me their help, I gave the place of honour to Eduard Fraenkel as the one to whom I owe more than to all the others. It was he who suggested that in writing this history I should pass straight from the Augustan age to the Italian Renaissance. After a long hesitation I adopted this suggestion, recognizing that neither my own inclination towards medieval literature nor my knowledge of the immense amount of research on medieval scholarship enabled me to produce a volume comparable to those on the ancient and modern ages. His untoward death in 1971 was a grievous loss to classical scholarship and a setback to my own work on its history.

It has been a great blessing that Mr. John Cordy of the Clarendon Press has been willing and able to continue his reading of the chapters in draft. I feel the deepest gratitude to his unfailing patience and competence. As before I must express my admiration for the skill and vigilance of the printers. In reading the proofs I have again enjoyed the help of my colleague Professor Max Treu in correcting a number of mistakes which had escaped my attention. I am also obliged to Mr. Ruprecht Volz, assistant on the Seminar of Nordisk philology in the university of Munich, for checking quotations and corrections.

I concluded the preface to the first volume by renewing an earlier dedication to my wife. With the deepest regret I can dedicate the new volume only to her memory, as she died after a long illness in February 1969, maintaining until the last day a passionate concern for the progress of my studies.

Munich   R.P.
January 1976

CONTENTS

ABBREVIATIONS   viii

PART ONE
THE RENEWAL OF CLASSICAL SCHOLARSHIP IN THE ITALIAN RENAISSANCE

I. Pre-humanism in Italy and the beginning of humanism: Petrarch and Boccaccio   3

II. The second and third generations: Salutati, Bruni, Niccoli, Poggio   25

III. Lorenzo Valla   35

IV. Politian   42

V. General achievements of scholarship in Italy and its spread into transalpine countries   47

PART TWO
HUMANISM AND SCHOLARSHIP IN THE NETHERLANDS AND IN GERMANY

VI. Devotio Moderna   69

VII. Erasmus of Rotterdam   71

VIII. Autour d'Érasme   82

PART THREE
FROM THE FRENCH RENAISSANCE TO THE GERMAN NEOHELLENISM

IX. Humanists and scholars of the French Renaissance   99

X. Classical scholarship in Holland and in post-renaissance France, Italy, and Germany   124

XI. Richard Bentley and classical scholarship in England   143

XII. Bentley's contemporaries and successors   159

PART FOUR
GERMAN NEOHELLENISM

XIII. Winckelmann, the initiator of Neohellenism   167

XIV. Friedrich August Wolf   173

XV. Wolf's younger contemporaries and pupils   178

XVI. The beginning of the nineteenth century. German Altertumswissenschaft from Niebuhr to Droysen   183

INDEXES   191

ABBREVIATIONS

AG	*Anecdota Graeca.*
AGGW	*Abhandlungen der Göttinger Gesellschaft der Wissenschaften.*
AJA	*American Journal of Archaeology.*
AJP	*American Journal of Philology.*
AL(G)	*Anthologia Lyrica (Graeca)*, ed. E. Diehl, 1925 ff.
APF	*Archiv für Papyrusforschung.*
Barwick, *Stoische Sprachlehre*	K. Barwick, 'Probleme der stoischen Sprachlehre und Rhetorik', *Abh. d. Sächs. Akad. d. Wissenschaften zu Leipzig, Phil.-hist. Kl.* 49.3 (1957).
BCH	*Bulletin de Correspondance Hellénique.*
Bursian	Bursians *Jahresbericht über die Fortschritte* der klass. Altertumswissenschaft.
Call.	Callimachus I, II, ed. R. Pfeiffer, 1949–53 (repr. 1965/6).
Cl. Phil.	*Classical Philology.*
Cl. Qu.	*Classical Quarterly.*
Cl. R.	*Classical Review.*
Colet, Opera	Colet, *Opera* ed. J. H. Lupton, 5 vols., 1867–76 (repr. 1965–69).
CMG	*Corpus medicorum Graecorum.*
DLZ	*Deutsche Literaturzeitung.*
DMG	*Deutsche Morgenländische Gesellschaft.*
Düring, 'Aristotle'	I. Düring, 'Aristotle in the ancient biographical tradition', *Studia Graeca et Latina Gothoburgensia* V (1957).
Erasmus, *Ep.*	Erasmus, *Opus epistularum*, ed. P. S. Allen vols. 1–12, 1906–58.
FGrHist.	*Die Fragmente der griechischen Historiker*, von F. Jacoby, 1923 ff.
FHG	*Fragmenta Historicorum Graecorum*, ed. C. Müller, 1841 ff.
GGA	*Göttingische Gelehrte Anzeigen.*
GGM	*Geographi Graeci minores*, ed. C. Müller, 1841.
GGN	*Nachrichten der Gesellschaft der Wissenschaften zu Göttingen.*
GL	*Grammatici Latini*, ed. H. Keil, 1855 ff.
GRF	*Grammaticae Romanae Fragmenta*, rec. H. Funaioli I (1907, repr. 1964).
Gr. Gr.	*Grammatici Graeci*, 1878–1910 (repr. 1965).
History [I]	R. Pfeiffer, *History of Classical Scholarship. From the Beginnings to the End of the Hellenistic Age* (Oxford 1968).
JHS	*Journal of Hellenic Studies.*
IMU	*Italia medioevale e umanistica.*
JHS	*Journal of Hellenic Studies.*
JRS	*Journal of Roman Studies.*
Kenyon, *Books and Readers*	F. G. Kenyon, *Books and Readers in ancient Greece and Rome* 2nd ed. (1951).
L–S	H. G. Liddell and R. Scott, *Greek-English Lexicon*. New edition by H. Stuart Jones, 1925–40.
Marrou	H.-I. Marrou, *A History of Education*, translated by G. R. Lamb (1956).
NJb.	Neue Jahrbücher für das klass. Altertum.
Nolhac	P. de Nolhac, *Pétrarque e l'humanisme*, 2 vols. (2nd ed. 1907, repr. 1959).
Pack²	R. A. Pack, *The Greek and Latin literary texts from Greco-Roman Egypt*, second revised and enlarged edition 1965.
Pasquali, *Storia*	G. Pasquali, *Storia della tradizione e critica del testo* (1934, repr. 1952).

Petrarca, *Prose*	F. Petrarca, *Prose*, a cura di G. Martellotti (and others), La letteratura italiana 7 (1955).
Philologia perennis	Rud. Pfeiffer, *Philologia perennis*. Festrede der Bayer. Akad. d. Wiss. (München, 1961).
*PLG*⁴	*Poetae Lyrici Graeci*, quartum ed. Th. Bergk, 1882.
PMG	*Poetae Melici Graeci*, ed. D. L. Page, 1962.
P. Oxy.	*Oxyrhynchus Papyri.*
PRIMI	Papiri della R. Università di Milano, vol. I, ed. A. Vogliano, 1937.
Prosatori	*Prosatori latini del Quattrocento.* A cura di E. Garin, La letteratura italiana 13 (1952).
PSI	*Papiri della Società Italiana.*
RE	Paulys *Real-Enzyklopädie der klassischen Altertumswissenschaft*, hg. v. Wissowa-Kroll-Mittelhaus, 1894 ff.
REG	*Revue des Études Grecques.*
Rh. M.	*Rheinisches Museum für Philologie.*
RML	Roscher, *Mythologisches Lexikon*, Suppl. 1921.
Rutherford, 'Annotation'	W. G. Rutherford, 'A Chapter in the History of Annotation', *Scholia Aristophanica*, III (1905).
Sabbadini, *Scoperte*	R. Sabbadini, *Le scoperte dei codici Latini e Greci nel secolo XIV e XV*, 2 vols (1905, repr. 1967).
Sandys	J. E. Sandys, *A History of Classical Scholarship*, 3 vols., 3rd ed. 1921 (1st ed. 1903).
SB	*Sitzungsberichte (Berl. Akad., Bayer. Akad., etc.).*
Schmidt, 'Pinakes'	F. Schmidt, '*Die Pinakes des Kallimachos*', *Klass.-philol. Studien I* (1922).
*SIG*³	*Sylloge inscriptionum Graecarum*, ed. W. Dittenberger, ed. tertia, 1915–24.
Steinthal,	H. Steinthal, *Geschichte der Sprachwissenschaft bei den Griechen und Römern mit besonderer Rücksicht auf die Logik*, 2 vols. 2. Aufl. 1890 (repr. 1961).
Susemihl	F. Susemihl, *Geschichte der griechischen Literatur in der Alexandrinerzeit*, 2 vols., 1891/2.
SVF	*Stoicorum Veterum Fragmenta*, ed. I. de Arnim, 1905 ff.
TAPA	*Transactions of the American Philological Association.*
*TGF*²	*Tragicorum Graecorum Fragmenta*, ed. A. Nauck, 2. ed., 1889.
Voigt, *Wiederbelebung*	Georg Voigt, *Die Wiederbelebung des classischen Alterthums oder das erste Jahrhundert des Humanismus* (1859).
Vors.	*Die Fragmente der Vorsokratiker*, von H. Diels. 6. Aufl. hg. v. W. Kranz, 1951–2.
Wendel, 'Buchbeschreibung'	C. Wendel, 'Die griechisch-römische Buchbeschreibung verglichen mit der des vorderen Orients', *Hallische Monographien* 3 (1949).
W. St.	*Wiener Studien.*
Wilkins, *Petrarch*	E. H. Wilkins, *Life of Petrarch*, 1961.

PART ONE

THE RENEWAL OF
CLASSICAL SCHOLARSHIP IN THE
ITALIAN RENAISSANCE

I

PRE-HUMANISM IN ITALY AND THE BEGINNING OF HUMANISM: PETRARCH AND BOCCACCIO

A GREAT Italian poet gave the original impetus to the revival of classical scholarship in modern times, Franciscus Petrarca (1304–74). This is not to claim him as the moving spirit of the Renaissance as a whole; even to discuss the general problem of its origins would be impossible here. But in our province, in the creation of a new method of approach to the literary heritage of the ancients, there can be no doubt that Petrarch was the protagonist. This irresistibly recalls the decisive part played by the early Hellenistic poets in the rise of scholarship in Alexandria.[1] Once again the revival of poetry, this time in the fourteenth century, preceded that of learning. Moreover, just as Hellenistic scholarship was indebted to the Peripatos to a certain degree, but opposed to it on fundamental issues, so also the scholarship of the Renaissance, though indebted to Aristotelian Scholasticism, to the traditional system of the *artes liberales*, and to the revived study of Roman Law, was yet in principle opposed to them.[2] In one important respect, however, Petrarch's position is not to be compared with that of Alexandrian scholar poets. He and his contemporaries and followers had more or less well-established texts of the ancient writers at their disposal, even if they were disfigured by bad copyists, and in some cases they had the benefit of the explanatory work of the ancient scholars. Servius' commentary on Virgil, for instance, contained precious relics of that exemplary Homeric scholarship , the development of which we traced in an earlier volume.[3]

In regarding Petrarch as the protagonist, we have not forgotten the existence of his predecessors in Italy, influential poets and scholars who were already in the later thirteenth and the early fourteenth centuries studying classical literature more intensively and striving to write a purer Latin themselves. Movements with the same tendency,

---

[1] See *History* [1] 87 ff.   [2] See below, p. 11.   [3] See *History* [1] 105 ff.

though different from each other in many respects, will be found in other countries, and they all may be called pre-humanistic.[1] Before Petrarch there was no humanism in the sense in which we shall presently define this protean term.

Petrarch lived, as he realized himself, between two ages: 'velut in confinio duorum populorum constitutus ac simul ante retroque prospiciens'.[3] Among his predecessors he seems to have duly acknowledged the literary merits of two Patavini, Lovato Lovati[4] (1241–1309) and Albertino Mussato[5] (1261–1329). Lovati's Latin poems were modelled on a few standard Roman poets, well known to the Middle Ages. Mussato tried to compose contemporary history in the manner of his fellow-countryman Livy and to produce a tragedy in the style of Seneca; he even wrote a metrical treatise on him, described the contents of his plays in the manner of the 'argumenta' (ὑποθέσεις) of late antiquity, and highly praised the moral character of the tragedies.[6] Mussato and Lovato struggled hard to get the facts right and to present them in a language nearer to ancient Latin than to the Latin of their predecessors; they went a little further than previous generations, but their achievement was modest and transitory, and it is incorrect to say that their circle and similar coteries in Venice, Vicenza, Milan, and Florence represented the beginning of humanism in Italy.[7]

Petrarch's[8] approach to the ancient writers was that of a powerful

[1] I avoid speaking of the 'dawn' of humanism, as it implies that there was previously the darkness of night.
[2] See below, p. 16.
[3] Petrarca, Rerum memorandarum libri, ed. Giuseppe Billanovich (Ed. nazionale delle Opere XIV, 1943) I 19.4.
[4] Ibid. II 61.   [5] Ibid. IV 118.2.
[6] A. C. Megas, 'The pre-humanistic circle of Padua (Lovato Lovati—Albertino Mussato) and the tragedies of Seneca', Ἀριστοτέλειον Πανεπιστήμιον Θεσσαλονίκης, Ἐπιστημονικὴ Ἐπετηρὶς Φιλοσοφικῆς Σχολῆς, Παράρτημα Ἀρ. 11 (1967) = 'Fourteenth-century Glosses and Commentaries on the Tragedy Octavia and on Seneca's tragedies in general, First part'. Summary pp. 229–33. This is a most welcome monograph on the earliest study of Seneca by the Paduans, based on new manuscript material, with an exhaustive bibliography and followed in 1969 by A. Mussato, Argumenta tragoediarum Senecae, Commentarii in L. A. Senecae tragoedias fragmenta nuper reperta, ed. A. C. Megas.
[7] Roberto Weiss, The Dawn of Humanism in Italy, Inaugural Lecture (London 1947) passim, esp. pp. 3, 21, and 'Il primo secolo dell umanesimo', Storia e Letteratura 27 (Roma 1949), Stuid e Testi, again on Padua, but also on Florence; a monograph 'Lovato Lovati', Italian Studies 6 (1951) 3–28 with bibliography.—G. Billanovich, I primi umanisti e le tradizioni dei classici Latini (Friburgo, 1953) with important new material and facsimiles.
[8] The best bibliographical survey of the editions of Petrarch's works is given by G. Foleno, 'Überlieferungsgeschichte der altitalienischen Literatur' in Geschichte der Textüberlieferung II (1964) 500–3 ('Giorgio Pasquali zum Gedächtnis'). The two Basle editions of the Opera 1554 (reprinted 1965) and 1581 are not yet completely superseded. In 1904 (the sixth centenary of Petrarch's birth) a 'Commissione per l'Edizione Nazionale delle Opere del Petrarca' was founded which so far has succeeded in publishing: Vol. I (1926) Africa; II (1964) De viris illustribus

poetically gifted personality, admiring classical form and penetrating
to the heart of the matter, and therefore successful for his own time
and for the future. A story related by Petrarch in old age[1] has become
famous. Francesco, born 1304 in Arezzo and taken in 1312 to the seat
of the papacy in Avignon, was destined by his father, an exiled Florentine lawyer, to the study of Roman civil law which he began in 1319 at
the university of Montpellier and finally gave up in 1326 in Bologna.[2]
The young student had little interest in the laws of debts and credits and
matter of that sort, but kept all the works of the Latin poets and of
Cicero that he could get hold of in a safe place for his favourite reading.
One day his father discovered them and flung them into the fire; when
Francesco cried out and burst into tears, the father saved two volumes
from the flames, a copy of Virgil and one described as Cicero's *Rhetoric*
(probably *De inventione*).[3] In telling the story Petrarch revealed what
he had felt as a boy: 'et illa quidem aetate nihil intelligere poteram,
sola me verborum dulcedo quaedam et sonoritas detinebat';[4] no one

---

1; x–xiii (1933–42) *Le Familiari*; xiv (1943) *Rerum memorabilium libri*; on the individual volumes
and their editors see below *passim*. A very welcome selection of prose-works and letters with
introduction and critical notes in: *La Letteratura Italiana*, Storia e Testi 7 (1955): *Prose* (1205 pp.);
the texts are revised by their respective editors, but often shortened. An indispensable guide
through the maze of editions and dates of the letters is E. H. Wilkins, *The Prose Letters of Petrarch*,
A manual (1951).

[1] *Lett. senil.* xvi 1 ed. Fracassetti, 1869/70 = Ed. Bas. 1581 p. 946 (cf. 1044) (to Luca della
Penna); W. Rüegg, *Cicero und der Humanismus. Formale Untersuchungen über Petrarca und Erasmus*
(Zürich 1946) pp. 8 ff. was right to start from the interpretation of this autobiographical
passage and to concentrate upon the peculiar 'form' of Petrarch's writings and its relation to
Cicero's style; but he consequently had to exclude any investigation of ideas or of scholarship.
Cf. also Karl Otto Apel, 'Die Idee der Sprache in der Tradition des Humanismus von Dante
bis Vico', *Archiv für Begriffsgeschichte* 8 (1963) 13 ff. These pages on Petrarch might be helpful,
which can hardly be said of the whole very large book (398 pp.). E. H. Wilkins, *Life of
Petrarch* (1961) tells the whole story year by year in a simple style without references; yet it is
not only an authoritative book based on a full knowledge of the sources and literature, but
also a sympathetic one with a rare sense of Petrarch's greatness and love of his personality.
P. de Nolhac, *Pétrarque et l'humanisme* (2 vols. 2nd ed. 1907, reprinted 1959) is fundamental for
our purpose; he had begun in 1890 to publish his researches and discoveries of Petrarch's
autographs and marginal notes on classical authors among the manuscripts of the
Bibliothèque Nationale de Paris and other libraries. In our day the most brilliant and
successful scholar in the same field is Giuseppe Billanovich (see below, p. 8 n. 3 on Livy and
passim); he celebrated the centenary of Nolhac's birth with a memoir 'Nolhac e Petrarca',
adding a list of books and articles in which Nolhac's great work is enlarged by new
discoveries, *Atti e Memorie della Accademia Petrarca di Lettere, Arti e Scienze*, n.s. 37, Anni
1958–64 (1965) 121–35. See also R. Sabbadini, *Le scoperte dei codici Latini e Greci ne' secoli
XIV e XV* (2 vols. 1905–14, reprinted 1967) and the list of his numerous other writings below,
p. 19 n. 4. E. Kessler, 'Petrarcas Philologie', *Petrarca, 1304–74 Beiträge zu Werk und Wirkung*,
hrsg. v. F. Schalk, (1974) 97 ff.

[2] G. G. Forni, 'F. Petrarca scolare a Bologna', *Atti e Memorie* (quoted above, n. 1 end)
pp. 83–96.

[3] P. de Nolhac, I[2] 221.1.

[4] Cf. Augustin. *Confess.* v 13, where as a boy listening to St. Ambrose, 'verbis eius
suspendebar intentus, rerum autem incuriosus . . . delectabar sermonis suavitate.'

for centuries had had an ear for the 'sweetness and sonority of words'.

'Questi son gli occhi della nostra lingua', he said of the two leading Romans in his lyric _Trionfo della Fama_ (III 21). In this striking metaphor a sort of emphasis is laid on 'nostra'; it is _our_ language they speak, they are our ancestors. Virgil had been rediscovered by Dante. Now Petrarch followed Virgil more closely in the resounding Virgilian hexameters of his epic poem _Africa_.[1] If ever a commentary[2] is written on this poem, it ought to show how far its language is indebted to Virgil, Livy, Cicero, and others. Petrarch avoids transcribing his ancient sources verbatim; with a rich vocabulary at his command he is able to give a phrase or a whole passage a new turn. So in _Afr._ II 544 he does not say 'et meritum maculare tuum' (as in Virg. _Aen._ x 851: 'tuum maculavi crimine nomen'), but with an intentional twist 'meritum vastare tuum'.[3] We shall find the same characteristic in his prose.[4]

The epic hero of the _Africa_ is Scipio Africanus;[5] he represents the greatness of the _Urbs aeterna_, which is in fact the theme of the whole poem. Petrarch transferred motives from Cicero's _Dream of Scipio_ (the younger) to his dreaming Scipio (the elder), unfolding in deliberately separated and differently styled parts of the first two books the past and the future of the glory of Rome. At the end in Rome Scipio's triumph and Ennius' coronation are celebrated. But the concluding book is also inspired by a tradition of a dream: the poet Ennius, Scipio's friend and—in Petrarch's poem—a companion of his African campaign, relates on their homeward journey to Rome a dream in which the shade of Homer had appeared and spoken to him, whose destiny it was to become 'alter Homerus'. Hesiod in the proem of his _Theogony_ had been the first to describe a poet's call by the Muses,[6] but the dream as the source of inspiration was introduced by Callimachus as the opening to his colloquy with the Muses. The motif of the dream was taken over by Ennius who replaced the godly Muses by the shade of the divine Homer (_Afr._ IX 159 ff.). In retelling this encounter Petrarch combined

---

[1] Petrarca, _L'Africa_, ed. N. Festa, 1926 (Ed. nazionale delle Opere 1), cf. E. Fraenkel, _Gnom._ 3 (1927) 485–94.

[2] Festa's critical edition of the text was without 'note explicative'; his _Saggio sull' Africa del Petrarca_ (1926), is no substitute for them. The 'Adnotata ad _Africae_ libros' which F. Corradini added to his edition in _Padova a Francesco Petrarca_ (1874) pp. 409–74 are still very useful.

[3] On the _varia lectio_ 'maculare—vastare' see the critical note in Festa's edition, who chose the correct reading for his text, but did not see the point. On this important problem of Petrarch's 'style' see Corradini, op. cit. pp. 100 f.

[4] See below, p. 15.

[5] Aldo S. Bernardo, _Petrarch, Scipio and the 'Africa'_ (Baltimore, Md. 1962).

[6] Hes. _Th._ 22 ff.; Call. fr. 2 Pf.; Enn. fr. A 5 ff. Vahlen; cf. A. Kambylis, _Die Dichterweihe und ihre Symbolik_, Bibliothek der klassischen Altertumswissenschaften (1965).

the few known fragments of Ennius with ideas of his own. Two characteristic passages may be mentioned: Homer's prophecy that 1,500 years after Ennius the great Scipio will again be duly praised by a poet 'Francisco cui nomen erit' (IX 233), and Ennius' admonition that poetry must be based on truth (IX 92 ff.): 'scripturum iecisse prius firmissima veri / fundamenta decet' ('sola quidem admiratio rerum / solus amor veri' II 453, the shade of Scipio's father had said in the first dream about the future poet).

This is not a conventional repetition of the phrase of the Hesiodic Muses, ἀληθέα γηρύσασθαι, but the assertion of a Petrarchan principle.[1] Petrarch built his *Africa* on the solid foundation of the third decade of Livy's history.[2] How eagerly he studied Livy is attested by his prose work *De viris illustribus*[2] as well as by the copy of Livy's text that contains his marginal notes. He started to compose the *viri illustres* shortly before the *Africa* and continued writing it side by side with the epic poem. Here too Scipio is celebrated with ecstatic admiration as the model of Roman virtue. Petrarch chose to give his Roman history (intended to run from Romulus to Trajan) the form of biographies of the great political and military men because for him only individuals counted; we find the same emphasis addressed to great literary men in his letters to the dead.[4] An example of the critical attention he paid to biographical facts and dates is his correction of traditional errors in the Life of Terence caused by Orosius' (IV 19.6) confusion of the comic poet Terentius Afer with Terentius Culleo; Petrarch had made the correction before it was confirmed by the discovery of Donatus' commentary on Terence with the *Vita* of Terence by Suetonius.[5]

Petrarch was the first to abandon the style of medieval chronicles, annals, and biographies and to combine excerpts and paraphrases of the original ancient sources—without quoting them verbatim—with occasional additions and corrections of his own. In this he was followed by the writers of Roman history for about three centuries, for until the later seventeenth century there was no question of replacing Livy by a newly written history.[6] He was endowed with an amazingly retentive

---

[1] Cf. E. Zinn, 'Wahrheit in Philologie und Dichtung', *Die Wissenschaften und die Wahrheit* (1966) pp. 134 ff., on the 'Wahrheitsproblem' especially in Roman and modern poetry.

[2] The same books of Livy were the source of Sil. Ital. *Punica*, rediscovered only in 1417 by Poggio and therefore unknown to Petrarch. The comparison of a plain versifier in his native idiom and a born poet in an ancient language is instructive also for Petrarch. On Sil. It. see M. v. Albrecht, *Silius Italicus* (1964) pp. 118 ff.

[3] *Opere*, ed. nazionale II (1964) ed. Martellotti, on chronology pp. ix ff.

[4] See below, p. 9 f.   [5] See Nolhac I 191, II 34.

[6] A. Momigliano, 'Contributo alla storia degli studi classici', *Storia e Letteratura* 47 (1955) 75.

memory. His life of Fabius Maximus, 35,[1] contains an excerpt from Livy XXII 29, where a famous Greek saying was quoted without the name of the author. Petrarch added 'notissimam illam Hesiodi poetae sententiam' (*Op.* 293 ff.), and it seems that he could only have done so by remembering the context of the quotation in a Latin version of Aristotle's *Nicomachean Ethics*.[2]

But Petrarch went much further than selecting, transcribing, and supplementing the subject-matter of Livy for his historical work; since the discovery of his own copy we know that he also tried to restore the text itself when he believed it to be corrupt.[3] He led the way to the revival of textual criticism; for his marginal notes are not merely illustrations and explanations as in other works. Given the opportunity of collating two manuscripts, he patiently recorded variant readings and skilfully emended a number of passages. No other scholar of his generation had either the good luck, or the talent to take advantage of it.[4] By an almost incredible chance it happened that the very copy owned by Petrarch passed a century later through the hands of Lorenzo Valla, who added his notes to those of Petrarch. Valla's criticism of Livy was known before the discovery of this manuscript, as his famous *Emendationes Livianae*[5] were published and even printed in the fifteenth century; but now we can see with our own eyes not only the hand of the great critic but the historical process itself in operation.

Looking back on this section which started with Virgil and Petrarch's *Africa* we realize that poetry began to play a new part in Italy, that Petrarch's poetical and scholarly work went side by side, and that he treated critically the text of that same source which he needed as much for the poem as for the prose work. It seems that Petrarch began to study Livy[6] early in his life, probably about 1318, twenty years before the start of the *Africa*; but whatever the historical sequence was, there could hardly be a more conspicuous example of the union of poetry and true scholarship. He was by no means only a littérateur.

---

[1] *De vir. ill.* ed. Martelletti, *Opere*, ed. naz. II 102.

[2] See the testimonia in Rzach's *ed. maior* of Hesiod; Aristot. *Eth. Nic.* I 2 p. 1095 b 9 ff. On Petrarch's acquaintance with the *Ethics* and their commentaries see Nolhac II 149 ff.; it was the first item in the catalogue of his library, ibid. I 42 f.

[3] G. Billanovich, 'Petrarch and the textual tradition of Livy', *Journal of the Warburg and Courtauld Institutes* 14 (1951) 137–208; cf. E. Fraenkel, *JRS* 42 (1952) 311. Billanovich started from an analysis of cod. Harl. 2493 in the British Museum; cf. Liv. ed. A. H. MacDonald, OCT v (1965) pp. viii, xix–xxv.

[4] Cf. also his critical notes on St. Jerome's translation of Eusebius, below p. 12 n. 1.

[5] See below, p. 36.

[6] Billanovich, op. cit. pp. 194 ff.

Examination of his own copy of Virgil[1] makes it evident that he lived in constant company not only with the poet, but also with his ancient interpreters, for Virgil's text is surrounded by Servius' commentary. It is not a pocket edition, but an extremely large and heavy manuscript, which he used to carry in his luggage when he travelled through France, Flanders, and Italy; and he covered its margins with innumerable and sometimes surprisingly learned notes.[2] He was able to point out the source of a scholion without the help of indexes or reference books. He remembered for instance (fol. 78 recto), that Servius' note on *Aeneid* II 254 'Phalanx lingua Macedonum legio' was taken from Livy XXXII 17.11, and on the note to *Aeneid* I 29 ff. 'Italus enim rex Siculorum profectus de Sicilia . . . ex nomine suo appellavit Italiam' he remarked (fol. 52 verso): 'secundum Tuchididem ut in octavo [i.e. VIII 328] "tum manus Ausoniae" ', which shows that he not only had this passage of Servius in his memory but could point beyond it even to the earlier Greek evidence (Thuc. VI 2.4). Modern editors have failed to take notice of these testimonia.[3] Petrarch's Codex Vergilianus is the most moving document of his immediate personal contact with the ancient Roman writers he loved.

Intimate as Petrarch's knowledge of Virgil and Livy was, others had laid the foundations. The rediscovery of Cicero was entirely his own. In 1333 he had found the speech *Pro Archia poeta* in Liège, but the really significant event came in 1345 when he rediscovered at Verona a manuscript of Cicero's *Epistulae ad Atticum*, *ad Quintum fratrem*, and *ad Brutum* (6–18); although tired and ill, he made a copy in his own hand,[4] and announced his exciting discovery to the world in the form of a letter addressed to Cicero himself.[5] One of the new features of the growing individualism of the Renaissance was a passionate interest in the personality of ancient writers which could be satisfied only by Latin not by Greek literature, and nothing was more fascinating to

[1] *F. Petrarcae Vergilianus Codex* [in bibliotheca Ambrosiana]; a complete reproduction ed. by Giovanni Galbiati, 1930; see the short enthusiastic review by E. Fraenkel, *Gnom.* 6 (1930) 552 f.

[2] Nolhac 1 140 ff.; G. Billanovich, *IMU* 3 (1960) 44.2 announced that Antonietta Testa is preparing an edition and a commentary on Petrarch's notes.

[3] I take these examples from E. Fraenkel's critical review of the Harvard edition of Servius' commentary, *JRS* 39 (1949) r47, as I could not find any better ones.

[4] *Ep. fam.* XXI 10.16 (ed. naz. vol. XIII) 'volumen quod ipse manu propria . . . scripsi, adversa valetudine'. It cannot be proved that the lost Veronensis is the archetype of the two families of the surviving manuscripts, nor is it likely that our oldest manuscript, Ambros. E 14 inf., has any connection with Petrarch; cf. Cic., *Letters to Atticus* ed. D. R. Shackleton Bailey 1 (1965) 77 ff.

[5] *Ep. fam.* XXIV 3; ibid. letters 4–12 were addressed to several ancient writers, 4 again to Cicero, and 12, the most extensive, to Homer.

Petrarch and his circle than this section of Cicero's correspondence. The letters to Atticus gave a true image of Cicero's personality, though Petrarch himself felt some disappointment when he recognized for the first time how much Cicero in time of civic unrest was distracted from his literary work by political acivity.[1]

Had Petrarch any knowledge of the other great section of Cicero's correspondence, the so-called *Epistulae ad familiares*?[2] Striking resemblances in contents and phraseology between Petrarch's letters of 1355 to Charles IV (*Ep. fam.* IX 4) and Cicero's letter to Caesar (Cic. *Fam.* VII 5) have been pointed out. Petrarch himself, however, never mentioned any discovery of the *Familiares*, nor did anyone else in his time. We know for certain only that- Salutati caused the Vercelli manuscript containing all sixteen books of the *Epistulae* to be copied for him in 1392.[3] It is not impossible that others had the chance of seeing parts of the collection before that date, but at present we simply cannot tell whether Petrarch actually did so.

Although much in Petrarch's letters was modelled on Seneca in the traditional manner when he treated moral topics, the more personal part became more and more influenced by the recently discovered Ciceronian style. Petrarch's letters were intended to be an embodiment of his life; he retained copies[4] for his 'immortality' and prepared collections at intervals during the fifties and sixties of the fourteenth century. In the hundreds of his letters still preserved there appears the most lovable trait in his nature, his desire for φιλία.[5] 'Never did any man form and cultivate a richer store of friendships'[6]—though one might perhaps make an exception of Erasmus. Mutual devotion often found the most moving expression. In Petrarch's life even his relations with his patrons all over Italy, ecclesiastical as well as secular ones, from whose benevolence and generosity he derived his livelihood, had a touch of honest affection. A strong feeling of friendship has remained characteristic of the true humanists and scholars of all ages.

The letters of his old age were crowned by a letter 'To Posterity';[7] no doubt, the model for this proud title was Ovid's autobiographical

---

[1.] On the different view of Salutati see below, p. 26.

[2] B. Kytzler, 'Petrarca, Cicero und Caesar', *Lebende Antike*, Symposion für R. Sühnel (1967) pp. 111 ff.

[3] Sabbadini, *Scoperte* II 214.

[4] *Ep. fam.* V 16.1–2, XVIII 7.8.

[5] *Ep. fam.* vol. IV = *Opere* ed. naz. XIII (1942) 375 f. Indice s.v. amicizia.

[6] Wilkins, *Life* p. 252.

[7] Reprinted by P. G. Ricci in *La Letteratura Italiana*, Storia e Testi 7 (1955) *Prose* pp. 2–19 'Posteritati' = *Senilium rerum libri*, XVIII.

letter (*Trist.* IV 10.2) 'Ille ego qui fuerim . . . quem legis ut noris, accipe posteritas.' Petrarch, having spoken to his Roman ancestors in a series of epistles, now spoke to his descendants; poet, sage, and scholar, the correspondent of popes and emperors, he stood in the centre as the dominating figure between past and future.

We are naturally inclined to give first place to the recovery of Cicero's lost works; but his philosophical treatises, which had not been unknown in the Middle Ages, were also treasured by Petrarch among his books, sometimes in more than one copy annotated by his own hand, and frequently quoted in many of his writings. It was the formal beauty of Latin poetry and prose that had first impressed him in his early youth, but far from becoming a mere lover of form, he was moved by a longing for true wisdom. He did not long, however, for the logic or metaphysics or natural sciences offered by the Aristotelian revival of the later Scholastic philosophy, but for knowledge of the human soul and human values. Against the Aristotle of the Scholastics he appealed to Cicero and to the greatest Ciceronian, St. Augustine. There are more than a thousand references to Augustine in Petrarch's writings.[1] The first book he could afford to buy in 1325 at Avignon was *De civitate Dei*,[2] and he always carried a tiny copy ('pugillare opusculum') of the *Confessions*,[3] a present of the Augustinian monk Dionigi; it was in his pocket even when he reached the summit of Mont Ventoux[4] where he opened it at random to be startled by the words of book X chapter 8. Some twelve years after buying the *Civitas Dei* Petrarch was able to acquire a part of Augustine's vast commentary on the *Psalms*[5] of which he later received a complete copy as a present from Boccaccio in the mid-1350s. The *Secretum*,[6] the most personal of Petrarch's works, is cast in the form of a dialogue between 'Franciscus' and 'Augustinus' about the Seven Deadly Sins; the Saint examines the penitent, who either repudiates the charges or pleads guilty. Even this Christian self-analysis is full of references to antiquity; for Petrarch made no distinction between the classics and the Church Fathers,[7] and collected their texts with equal zeal. Among the Fathers he particularly venerated

---

[1] P. P. Gerosa, *Umanesimo Cristiano del Petrarca. Influenza Agostiniana* (1966).

[2] G. Billanovich, 'Nella biblioteca del Petrarca', *IMU* 3 (1960) 2.

[3] P. Courcelle, *Les Confessions de St. Augustin dans la tradition littéraire*. Antécédents et Posterité (1963) pp. 329–51, 'Un Humaniste épris de confessions: Pétrarque'.

[4] *Ep. fam.* IV 1.

[5] *Enarrationes in Psalmos*, see Billanovich, op. cit. (above, n. 2) pp. 5 ff.

[6] *Prose* pp. 22–215, 'De secreto conflictu curarum mearum' ed. Carrara (see critical notes pp. 1162 f.).

[7] On 'classics' see below, p. 84 n. 4.

St. Ambrose and St. Jerome,[1] after Augustine who always held the first place in his affection. He strove to follow the common moral teaching of the ancient writers, whether Academic, Stoic, or Christian; for his literary studies he had only one aim, as he stated in one of his latest writings (after 1363): 'Tu scis, Domine, quod ex literis . . . nihil amplius quaesivi quam ut bonus fierem. Non quod id literas aut . . . omnino aliquem, nisi te unum facere posse confiderem, sed quod per literas quo tendebam iter honestius ac certius simulque iucundius existimarem, te duce, non alio. . . . Nunquam . . . tam gloriae cupidus fui . . . quin maluerim bonus esse quam doctus'.[2]

The imaginative enthusiasm of Petrarch was not limited to the few leading writers; he began to search the libraries for all the literary treasures of Roman antiquity which he regarded as his own ancestry.[3] It was a fortunate chance that he and other exiled Italians lived in the south of France and could without too much difficulty reach the libraries of the French monasteries and especially those of the great cathedrals. The manuscripts hidden there were made accessible to him and to other individual scholars by means of careful copies.[4] In the end nearly all the classical Latin texts known in his time had been collected, read, and more or less fully annotated in their margins by Petrarch himself; he was the first man of letters in modern times to build up a private library of this kind.[5] But not only that; in 1362, with the ancient libraries of Alexandria and Rome in mind, he had the highly original idea of bequeathing his own (his 'daughter', as he called it) to the Republic of Venice as the nucleus of a future public library.[6] But the eventual fate of Petrarch's beloved books was quite different. In his will of 1370 they were not mentioned; they were apparently removed to Arquà and, when he died there in 1374, they suffered the dispersal which he had tried to avoid. Still, a substantial part of the library

---

[1] G. Billanovich, 'Un nuovo esempio delle scoperte e delle letture del Petrarca, L' "Eusebio-Girolamo-PseudoProspero" ', *Schriften und Vorträge des Petrarca-Instituts Köln* 3 (1954). One of the most astonishing examples of Petrarch's zeal, learning, and memory is the notes in his copy of Eusebius' *Chronicon*, translated by St. Jerome (published on pp. 26–50), and Billanovich (p. 14) was fully justified in declaring: 'Il Petrarca ebbe nella storia della filologia un' importanza eguale, o persino maggiore, di quella che ebbe nella storia della poesia.'

[2] 'De sui ipsius et multorum ignorantia' ed. P. G. Ricci in *Prose* 716 (cf. pp. 1173 ff.).

[3] G. Billanovich, 'Il Petrarca e i classici', *Studi Petrarcheschi* 7 (1961) 24.

[4] Cf. the very useful indexes to the letters in *Opere*, ed. naz. vol. XIII (1942): pp. 349 Libri di Petrarca, 384 copisti, 401 f. libri degli antichi perduti, ritrovati, etc., 419 s.v. scrittore: antichi scrittori classici, 423 studi dell'antichità.

[5] *Ep. fam.* III 18.2 'libris satiari nequeo, et habeo plures forte quam oportet . . . quaerendi successus avaritiae calcar est.' See Nolhac I 163 ff. poets, II 1 ff. prose authors.

[6] *Petrarch's Testament*, edited and translated by Theodor E. Mommsen (1957) pp. 42–50 of the introduction.

survived, notably that which went via Pavia to Paris,[1] and the scrupulous detective work of modern scholars[2] has identified many of the other volumes scattered over Western Europe.

Petrarch had an ear for 'the sweetness and sonority' of ancient Latin, as he himself confessed. A feeling began to spread that beauty of form should be matched by beauty of script. In searching for classical manuscripts, Petrarch and his followers came across the earlier medieval script,[3] the Carolingian minuscule, which seemed to them of venerable antiquity and beauty, and therefore to call for revival. We assumed[4] that the aesthetic sense of the scholar poets of the third century B.C. was responsible for a characteristic change of script. Now again a slow scribal reform started, and signs of transition from the so-called Gothic[5] to the humanistic script can be observed in Petrarch's copies.[6] In the writing of vernacular texts and even of modern Latin verses there was no change; but for the transcription of classical texts and for the scholar the use of the 'littera antiqua', which Petrarch called 'castigata et clara',[7] became more or less obligatory. It is likely that after Salutati's experiments the new style of handwriting was fixed by Poggio.[8]

Petrarch had an essentially Latin mind, and the movement he initiated was centred on Latin for generations. Greek language and literature were little more than a vision, a dream-world. In the handsome codex of Suetonius[9] used by Petrarch as a working copy on his travels the scribe had left blank spaces for quotations and tags in Greek, which seem to have been filled in by Petrarch himself in rather

---

[1] E. Pellégrin, *La Bibliothèque des Visconti et des Sforza* (1955); cf. Martellotti (above p. 7 n. 3) p. xv n. 6.

[2] See the many references to P. de Nolhac and G. Billanovich.

[3] About one-third of Petrarch's and of Salutati's manuscripts, known at present, are of the ninth to twelfth centuries.

[4] See *History* [1] 103.

[5] On the coining of the term 'Gothic' by Valla see below, p. 35.

[6] J. Wardrop, *The Script of Humanism. Some Aspects of Humanistic Script 1460–1560* (1963) pp. 5 f. and pl. I. These lectures were delivered in 1953 and published after the author's death; the first chapter on the rise of the humanistic cursive was written long before the appearance of Ullman's book, see below, n. 8. Cf. H. Hunger in *Geschichte der Textüberlieferung* I (1961) 143 'Gothico-Antiqua' . . . 'Petrarca—Schrift'. Excellent plates of ancient, medieval, and humanistic script in Giuseppe Turrini, *Millennium scriptorii Veronensis dal IV al XV secolo* (1967).

[7] *Ep. fam.* XXIII 19.8 'non vaga quidem ac luxurianti litera (qualis est scriptorum . . . nostri temporis . . .), sed alia quadam castigata et clara'.

[8] B. L. Ullman, 'The Origin and Development of Humanistic Script', *Storia e Letteratura* 79 (1960) 21 ff.; pl. 4 Petrarch's script.

[9] R. W. Hunt, 'A Manuscript from the Library of Petrarch' (Oxford, Exeter College 186), *Times Literary Supplement* (23 Sept. 1960) p. 619; G. Billanovich, 'Nella biblioteca del Petrarca', *IMU* 3 (1960) 28–58 (Un altro Suetonio di Petrarca).

awkward Greek letters. We can see him here struggling with the very elements of the Greek alphabet, but he never gave up the struggle until the last day of his life. In Cicero as well as Augustine he found Plato often quoted and highly praised; writings on Virgil, Macrobius and Servius' commentary, persistently referred him to Homer, whose shade (as we have seen) introduced an important passage of his *Africa*.[1] The arrogant Aristotelians, the Averroists of his time, knew nothing of Plato, while Petrarch could proudly assert 'sedecim vel eo amplius Platonis libros domi habeo',[2] namely in his library at Vaucluse. His annotated copy of the Latin *Timaeus*[3] with a commentary by Chalcidius is still extant, and he is known to have possessed the *Phaedo* in the Latin version of Henricus Aristippus.[4] In his *Rerum memorandarum libri* I 25 he expressed his profound reverence for the 'philosophorum princeps'. Of Plato's works he saw more in the library of the Basilian monk and bishop Barlaam,[5] who started (probably in 1342) to teach him a little Greek and even introduced him to Homer. In 1354 Nicholas Sigeros,[6] the envoy of the Byzantine emperor to the papal court in Avignon, presented him with a copy of the *Iliad*,[7] which he enthusiastically embraced; but he had to confess: 'Homerus tuus apud me mutus . . . quam cupide te audirem.'[8] It was not until four or five years later that Petrarch could actually hear Homer speak, in the literal Latin translation of Leonzio Pilato, a Calabrian like Barlaam, whose mother language was Greek; he had already Latinized five books of the *Iliad* before he was persuaded by Petrarch and Boccaccio in Florence to make a complete translation of the two Homeric poems.[9] In a beautiful transcript of Pilato's translation the old Petrarch illuminated with trembling hand the whole *Iliad* and the *Odyssey* up to β 242 (Par. 7880); a note of Pier Candido Decembrio tells us[10] that he died on 23 July 1374 while annotating this volume.

Petrarch's attempts to learn Greek were abortive; but everyone

[1] See above, p. 6.   [2] 'De ignorantia' ed. Ricci in *Prose* p. 756; cf. Nolhac II[2] 134 ff.
[3] R. Klibansky, *The Continuity of the Platonic Tradition* (1939) p. 30.
[4] L. Minio-Paluello, 'Il Fedone Latino con note autografe del Petrarca', *Atti della Accademia dei Lincei* 1949, Ser. VIII, Rendiconti, Classe di scienze morali, storiche e filologiche, IV 107 ff.
[5] On the library of Barlaam, who possessed Eur. cod. Laur. XXXII 2, see B. Hemmerdinger, *REG* 69 (1956) 434 f.
[6] A. Pertusi, 'Leonzio Pilato fra Petrarca e Boccaccio', *Civiltà Veneziana, Studi* 16 (1964) 43–72.
[7] Petrarch had asked him also for copies of Hesiod, Herodotus, and Euripides.
[8] *Fam.* XVIII 2 (10 Jan. 1354); Pertusi, op. cit. pp. 65 ff. identified Petrarch's Homer with the cod. Ambros. gr. 198 inf.
[9] An anonymous writer made another translation of the *Odyssey* into Latin prose before 1398, see Pertusi, op. cit. pp. 53 ff.
[10] Nolhac II[2] 167.

could feel how ardently he longed to know the Greek background of Roman literature. So his attempts stirred others; this is characteristic of Petrarch in general: even when his own efforts were not successful, they had an inspiring effect on later generations.

Petrarch had learned from Cicero that the Romans regarded the Greeks not only as literary models, but as the 'most human people', the *genus humanissimum* who had set an example of human culture (παιδεία), valid for all people and for all time. This new Roman concept called *humanitas* could be found everywhere in Cicero's writings;[1] Petrarch, however, used the word sparingly. But there is a striking sentence in the dedicatory letter to *De vita solitaria*,[2] addressed to bishop Philip of Cavaillon in 1366, twenty years after the first draft of the book: 'perniciosum quoque et varium et infidum et anceps et ferox et cruentum animal est homo, nisi, quod rarum Dei munus est, humanitatem induere feritatemque deponere . . . didicerit.' When I first came across the phrase 'humanitatem induere feritatemque deponere',[3] I thought it must be borrowed from an ancient source, presumably from Cicero, as 'humanitas', set in opposition to 'feritas', occurs there.[4] I was wrong; the phrase as a whole is not borrowed from elsewhere, but coined by Petrarch himself who ingeniously combined Cic. *ad Att.* XIII 2.1 'humanitatem omnem exuimus' (cf. *Lig.* 14) with Ovid. *fast.* IV 103 'deposita . . . feritate' (sc. *taurus*). This sort of variation and combination is exactly his style in poetry and prose.[5] Here and in a few other passages[6] 'humanitas' means human feeling, a compassionate attitude to one's fellow men, φιλανθρωπία. But Petrarch was convinced that the *literae* he cultivated paved the way to moral values and true wisdom; there was therefore a definite relation between *literae* and *humanitas*. Petrarch used to speak of his own love and knowledge of 'vetustas'.[7] In his letter to 'Posterity'[8] he confessed: 'Incubui unice, inter multa, ad notitiam vetustatis',[9] and he was well aware that his

---

[1] See *Thes. Linguae Lat.* s.v. humanitas; *Humanitas Erasmiana* (1931) pp. 2 ff.; F. Klingner, 'Humanität und Humanitas', *Römische Geisteswelt* 5. Aufl. (1965) 704 ff., esp. 718 ff. and notes and 741 ff. on Cicero.

[2] Ed. G. Martellotti in *Prose* p. 294.1; cf. pp. 1166 ff.

[3] E. Arnold referred me then to G. Paparelli, 'Feritas, humanitas, divinitas, le componenti dell'Umanesimo', *Biblioteca di cultura contemporanea* 68 (1960) 31–47, on the sentence quoted above; but Paparelli did not see its relation to Cicero and Ovid.

[4] Klingner, op. cit. (above n. 1) p. 743.66 'contrasts to humanitas'.

[5] See above, p. 6.

[6] *Ep. fam.* VI 3.3 (with reference to Ter. *Haut.* 11.25), *Ep. fam.* XII 2.28, *Sen.* XIII 15.

[7] See above, p. 9.   [8] *Prose* p. 6.9.

[9] Cf. 'Invectivae contra medicum quendam' (*Opera*, ed. Basil, 1554, reprinted 1965, p. 1199): 'nihil mihi carius quam vetustas ipsa, cuius venerator nostra aetate nisi fallor nemo †inde maior fuit.'

passionate love was infectious: 'ad haec nostra studia, multis neglecta saeculis, multorum me ingenia per Italiam excitasse et fortasse longius Italia.'¹ So Petrarch wrote in 1373 shortly before his death in a very moving letter to Boccaccio in which he expounded to him the reasons why one should not interrupt one's studies because of old age.² The strong belief in the lasting effect of his work expressed in this prophetic sentence was confirmed by his most faithful admirers in the next generation. In Leonardo Bruni's *Dialogi*³ of the year 1401 Niccolò Niccoli says about Petrarch: 'Hic vir studia humanitatis, quae iam extincta erant, reparavit'; a few months later Salutati used the same expression 'studia humanitatis'.⁴ It seems to have become established in this sense in the lively discussions of the learned Florentine circle. We have often been told that humanism arose from the social and political conditions of the consolidated new Italian city states; and it is true that these conditions became more and more favourable to the development and diffusion of Petrarch's ideas. These ideas, however, originated from his own mind; they did not spring from the spirit of the society of his time of which he always spoke with contempt ('mihi semper aetas ista displicuit'⁵). It was because his studies of antiquity were shortly afterwards termed 'studia humanitatis' by the leading members of the Florentine circle that the critical scholarship which he recreated became amalgamated with the concept of *humanitas* for the whole future, as did no other branch of scholarship. This union, as we shall see, involved many problems in the course of time; it was due, as we have tried to explain, to the personal impulse of an original poetical genius.

¹ *Senil. rer.* l. xvii 2 = *Prose* p. 1144.14.

² *Prose* pp. 1134 ff., 1156.24 f. 'An tu vero forsitan non Ecclesiasticum illum audisti: "cum consumaverit homo tunc incipiet, et cum quieverit tunc operabitur."' ' Martellotti quotes in his apparatus: Eccles. 18: 6 '. . . cum quieverit, aporiabitur.' This note is misleading in that it implies that 'aporiabitur' is the traditional text of the Vulgate. It actually is the text in our editions from 1598, the date of the editio Clementina, up to *Biblia sacra*, Vulgatae editionis nova editio, 1955, p. 917; but nearly all of the 30 or so manuscripts read 'operabitur', as the new critical edition shows (not yet at Martellotti's disposal): *Biblia sacra iuxta Lat. vulg. versionem* xii (1964) Sirach 18: 6, and as the critical edition of the *Vetus Latina* (now in preparation at Beuron) will confirm. I am very much obliged to Dr. W. Buchwald of the Munich *Thesaurus Linguae Latinae* for his kind help.—I shall not decide whether 'aporiabitur', which corresponds to the Greek text ἀπορήσεται and is preserved also in Ambros. *Expos. psalmi* 118 (*CSEL* 62) serm. 8.17.3, or 'operabitur' is the original reading of the Vulgate. It is sufficient to state that Petrarch quoted the text current in the medieval manuscripts.

³ L. Bruni, 'Ad Petrum Paulum Histrum dialogus', hg. von Th. Klette, *Beiträge zu Geschichte und Literatur der italienischen Gelehrtenrenaissance* 2 (1889) 80. W. Brecht in K. Brandi, *Das Werden der Renaissance* (1908) pp. 22 ff., was the first to refer to this important passage. See below, p. 30.

⁴ C. Salutati, *Epist.* iii p. 599 ed. Novati: 'Erit aliquis studiis humanitatis locus.'

⁵ 'To Posterity', *Prose* 6. 10 etc.

As Petrarch was celebrated shortly after his death for having been the first to restore the 'studia humanitatis' we can properly apply the modern term 'humanism'[1] to the age of this restoration. In Petrarch's own phraseology 'humanitas' meant φιλανθρωπία, but it was used by Salutati and Bruni to describe his literary studies. Similarly in the nineteenth century the German neologism 'Humanismus' was coined for an educational theory (1808),[2] then used for the cultural movement opposed to 'Scholasticism' (1841),[3] and finally (1859) applied to the specific period of the revival of classical studies[4] by Georg Voigt, whose book on that period bore the subtitle 'the first century of humanism'.[5] For a century this book has remained the standard work on its subject, consulted by every student of the revival of classical antiquity in Italy.[6] It is a sober, solid, and readable collection of material, and though antiquated in many respects, can even now provide useful information. But it also has its dangers in so far as Voigt ventures to express his own opinions on the tendencies and achievements of the Italian humanists. Looking at them from the point of view of German Protestant liberalism, he called the literary, educational, and religious aims of the Italians childish and fantastic, and regretted their lack of Teutonic soul. It is a strange paradox that a man of his outlook should have felt impelled to make those most detailed and comprehensive researches without any real understanding and sympathy; the same could be said about his long monograph on Aeneas Sylvius Piccolomini.[7]

---

[1] A survey of modern studies of humanism is desirable; it cannot be squeezed into the text or the notes. Access to the original texts is made easier by the collection of extracts in *The Renaissance Debate*, edited by Denys Hay (1965) with short introduction and additions. Hay's survey is of course not confined to the scope of my book on scholarship. W. K. Ferguson, *The Renaissance in Historical Thought. Five centuries of interpretation* (Cambridge, Mass. 1948), traced all the variations in conception and interpretation of the Renaissance; see especially pp. 386 ff.

[2] F. Niethammer, *Der Streit des Philanthropinismus und Humanismus in der Theorie des Erziehungsunterrichts unserer Zeit* (1808). In this title the first -*ismus* (derived from Basedow's 'Philanthropinum') seems to have provoked the second -*ismus*; this new formation was accepted by every European language in the course of the nineteenth century. It is not superfluous to recall the *origin* of this much discussed word, as it is so often forgotten. See W. Rüegg, *Cicero und der Humanismus* (1946) pp. 2 ff. and W. Kaegi, *Humanismus der Gegenwart* (1959) pp. 24 ff., 58 ff. (on 'humanista').

[3] K. Hagen, *Deutschlands literarische und religiöse Verhältnisse im Reformationszeitalter* 1 (1841); only the second edition (1868) was available to me, see ch. I, p. 39 'Repräsentanten des Humanismus', p. 79 'Annäherung an den Humanismus', etc.

[4] Georg Voigt, *Die Wiederbelebung des classischen Alterthums oder das erste Jahrhundert des Humanismus*, 1st ed. 1859.

[5] 4th ed. (unveränderter Nachdruck der ... dritten Auflage), Berlin 1960, W. de Gruyter.

[6] See J. A. Symonds, *The Renaissance in Italy*, vol. II 'The Revival of Learning' (1877, 2nd ed. 1882); J. E. Sandys, *Harvard Lectures on the Revival of Learning* (1905).

[7] G. Voigt, *Enea Silvio Piccolomini als Papst Pius der Zweite und sein Zeitalter*, 3 vols. (1856–63).

One year after Voigt's *Wiederbelebung* there appeared one of the most brilliant and influential works of scholarship written in the last century, a masterpiece of historical reconstruction in perfect German prose: Jacob Burckhardt's *Die Kultur der Renaissance in Italien*. The first edition[1] was modestly called 'Ein Versuch' by the author. The relatively small section 'Die Wiedererweckung des Altertums' is not concerned with the history of learning[2] in Italy, but with the reproduction of antiquity in literature and life, the amalgamation of the reborn spirit of Roman antiquity with the Italian national character ('mit dem italienischen Volksgeist'), for which he felt a spontaneous and permanent affection. Jules Michelet is supposed to have been the first modern historian to apply the comprehensive term 'Renaissance' to the whole epoch[3]; but though Burckhardt did not coin it, it was he who made it popular, and his essay was the starting-point of all the subsequent discussions about the beginning and the concept of the 'Renaissance'. A fervent opponent of Burckhardt was Konrad Burdach,[4] who believed he recognized the true origin of the new epoch in Cola di Rienzo's fancies and visions, ecstasies and ideas of the rebirth of Rome (with Rienzo as tribune). But if we keep to the traditional division of the historical periods, the wealth of material published and interpreted by Burdach and his collaborators shows Rienzo's pseudo-religious mysticism as characteristic of the troublesome dissolution of the Middle Ages, not as heralding a new age. After Burckhardt the most valuable new treatment of the main problems of the Renaissance was E. Walser's *Gesammelte Studien zur Geistesgeschichte der Renaissance* (1932);[5] they were the work not of a historian, but of a professor of Romance languages in Basle who had a genuine affection for everything Italian and had, as biographer of

---

[1] Only this edition (Bâle 1860) and the second edition (1869) were authentic; in the following ten editions (in two volumes, 1877–1919) notes and excursuses were added and constantly enlarged, until the thirteenth edition returned to the original, which is reprinted with an important introduction by W. Kaegi in vol. v (1930) of the Gesamt-Ausgabe in 14 volumes (Stuttgart 1929–34).

[2] He regretted not to be able to refer to 'eine gute und ausführliche Geschichte der Philologie'.

[3] *Histoire de la France* vol. vii (1855) quoted by Burckhardt p. 219.1. Perhaps a collection under the title *Le Moyen Âge et la Renaissance*, . . . Direct. litt. Paul Lacroix, 5 vols. (Paris 1848–51) can claim priority; it was, as far as I can see, not mentioned in the whole discussion. On the history of the word see B. L. Ullman, 'Renaissance, the word and the underlying concept', in *Studies in the Italian Renaissance* (1955) pp. 1 ff.

[4] *Vom Mittelalter zur Reformation*, 11 vols. (1893–1937).

[5] W. Kaegi collected and edited the studies after the early death of the author, with an extensive introduction 'Über die Renaissanceforschung E. Walsers'. The collection contains also the six lectures delivered at Cambridge in 1926, 'Human and artistic problems of the Italian Renaissance', in a German translation, pp. 211–326; see especially 'Das antike Ideal' and 'Homo et Humanitas' pp. 308 ff.

Poggio, acquired the most intimate knowledge of Renaissance literature. He protested against the simplifications and exaggerations of Burckhardt's followers, notably against the general assertion that the Renaissance was an irreligious, pagan, and enlightened period in sharp contrast to the Middle Ages, and he stressed the need for a new careful interpretation of the so-called anti-medieval utterances against the condition of the Church and against scholastic philosophy. W. Dilthey's essays *Weltanschauung und Analyse des Menschen seit Renaissance und Reformation*, written from 1891 onwards,[1] suggest many good general ideas. Indeed, as we have seen in the paragraphs on Petrarch and shall see throughout our account of the period between Petrarch and Erasmus, the Renaissance was not anti-Catholic, un-Christian, atheistic, but a period in which men were trying to find a new, more personal piety and new expressions of religious thought. There was a slow change inside the Church, aiming at reform, not a negation of the past. These and similar notions would probably have stood in the centre of Walser's great plan of a 'Geistesgeschichte der Renaissance'; it would have been a substantial improvement on the concluding chapter of the *Kultur der Renaissance*.

The studies of two Italian scholars serve as critical supplements to Burckhardt's work: G. Toffanin took a strong line against the supposed 'paganism' of the Italian Renaissance in his books on humanism,[2] and E. Garin ably filled the gap left by the absence of a section in Burckhardt on the philosophy of the Renaissance.[3] For our purpose not only are the often quoted *Scoperte* of R. Sabbadini on the discovery of classical manuscripts indispensable, but also his many other writings on Italian humanism, unfortunately not yet accessible in a collection of reprints.[4] Humanistic manuscripts of the Renaissance in Italian and other literatures have been listed by P. O. Kristeller.[5] The *Studies in the Italian Renaissance* (1955) of B. L. Ullman, a connoisseur of classical,

---

[1] W. Dilthey, *Gesammelte Schriften* II[2] (1921) 19 ff., 322–6, and *passim*.

[2] G. Toffanin, *Storia dell' umanesimo dal XIII al XVI secolo*, 2nd ed. (1940) with bibliography pp. 369–88 (Nuova edizione 1964, 4 vols.). It is not easy to follow his sometimes perverse train of thought; when he postpones 'nascità della filologia' until the end of the sixteenth century (pp. 329 ff.), he is refuted by the evidence now available from Petrarch to Valla.

[3] E. Garin, L' umanesimo italiano. Filosofia e vita civile nel Rinascimento', *Biblioteca di cultura moderna*, No. 493, 2nd ed. 1958 (see especially pp. 11 f., 64, 82 ff.).

[4] See the bibliography of his books and papers from 1878 to 1932 in *Fontes ambrosiani* II (1933) and in R. Sabbadini, *Storia e critica di testi latini* (2nd ed. 1971), bibliography pp. xi–xli from 1873 to 1936; but no collected reprints. It is promised, however, on p. x of the second edition of the *Storia e critica* that *Opere minori*, 'alcuni volumi', will follow soon in the collection Medioevo e Umanesimo.

[5] P. O. Kristeller, *Iter Italicum* I (1964), II (1967); cf. G. Billanovich's authoritative review in *Gnomon* 42 (1970) 217 ff.

medieval, and Renaissance literature and palaeography, contain much new material, and W. Rüegg who ,started from stylistic researches on Petrarch[1] has continued and enlarged his studies over the whole period.[2]

Giovanni Boccaccio (1313–75) was Petrarch's junior by nine years and died a few months after him,[3] enthusiastically devoted to him and eager to promote and spread his fame, especially in Florence.[4] Petrarch, for his part, could not fail to recognize and appreciate the literary and scholarly qualities of his follower, though Boccaccio's nature, interests, and achievements differed widely from his own. Boccaccio was born in Paris, the illegitimate son of a French woman and an Italian merchant from Certaldo; he grew up in Naples,-failed as a merchant and as a student of canon law, but met with success as the narrator of the short stories, partly 'frivolous and partly moralizing, assembled under the famous title *Decameron*. Perfect artistry reached its peak in the serious stories of the tenth day which impressed Petrarch so much that he even translated the concluding story of Griselda into Latin (*De insigni obedientia et fide uxoria*), discussing the whole matter in his correspondence with Boccaccio. We are told that a visit to the tomb of Virgil at Naples awakened Boccaccio's lasting enthusiasm for ancient poetry.[5] But in spite of his sincere love of Virgil, Dante, and Petrarch, and although he produced a fair number of lyric and epic poems in his native tongue, he was a realist by nature, occupying himself with things rather than words.

This is obvious in his learned collections of ancient mythological, historical, and geographical material. The *Genealogie* [*sic*] *deorum gentilium*[6] is based on the so-called *Mythographus Vaticanus* III of the later Middle Ages,[7] but owes much to the help of the Greek Calabrian,

[1] See above, p. 5 n. 1.

[2] See especially.the Züricher Ringvorlesungen in *Erasmus-Bibliothek* 'Das Trecento' (1960) pp. 139 ff. and 'Das Erbe der Antike' (1963) pp. 95 ff. with further references. The main problems of humanism and Italian Renaissance are rediscussed by Ch. Trinkaus, *In Our Image and Likeness. Humanity and divinity in Italian humanist thought*, 2 vols. (1970).

[3] A short biography by E. Walser in *Gesammelte Studien zur Geistesgeschichte der Renaissance* (1932) pp. 38 ff. On the complicated tradition of his numerous writings, particularly on the autographs, see Pasquali, *Storia* pp. 443 ff. and G. Folena in *Geschichte der Textüberlieferung* II (1964) 503 ff.—'Epistularum quae supersunt' in *Opere Latine minori* ed. A. F. Massèra (Bari 1928) pp. 109–227.

[4] G. Billanovich, 'Petrarca letterato I. Lo scrittoio del Petrarca', *Storia e Letteratura* 16 (1947) 57–294: 'Il piu' grande discepolo' [i.e. Boccaccio].

[5] F. Villani, *De civitatis Florentiae famosis civibus*, ed. G. C. Galletti (1847) p. 17.

[6] The recent edition by V. Romano (Bari 1951, Scrittori d'Italia 200–1) presents the text of Boccaccio's personal copy (Cod. Laur. plut. LII 9).

[7] O. Gruppe, 'Geschichte der klassischen Mythologie und Religionsgeschichte', *RML*, Supplement (1921) 22 ff., H. Liebeschütz, 'Fulgentius Metaforalis', *Studien der Bibliothek*

Leonzio Pilato,[1] and his notes on the Latin translation of the Homeric poems. Its contents were not only an inexhaustible source for students of mythology, but also an inspiration for Renaissance poets[2] and artists until the middle of the sixteenth century, when it was more or less superseded by the *Mythologia* of Natalis Comes (1551).[3] Boccaccio's book preserved late classical and medieval explanations of myths in the Stoic[4] allegorical tradition. In his comparisons of ancient myths with Christian legends there seems to be a slight shifting of emphasis to the advantage of the classics, inconceivable in pre-Renaissance times.

Errors in a work popular through two centuries are occasionally of curious consequence. In the first book, for instance, the genealogy of the gods is headed by the god 'Demogorgon'[5] as father of Uranus. The mysterious name which sounds so very archaic found its way into Italian poetry[6] from Boiardo's *Orlando innamorato* and Ariosto's *Cinque canti* (not *Orlando furioso*) to Carducci and D'Annunzio, into French literature from Arnoul Greban's *Mystère de la Passion* to Rabelais and Voltaire, and with even greater vigour into the masterpieces of English poetry[7] from the sixteenth to the nineteenth century; Spenser in his *Faery Queene* seems to have been the first to introduce what Milton later called 'the dreaded name of Demogorgon'.[8] Demogorgon, however, is no relative of the recently discovered formidable Hurrian

Warburg 4 (1926) 20 f. J. Seznec, 'The Survival of the Pagan Gods', *Bollingen Series* 38 (1953) 220 ff.

[1] See Pertusi (above, p. 14 n. 6) pp. 295 ff.; he prefers the spelling 'genologie'.

[2] It was consulted by Chaucer not long after Boccaccio's death, see Chaucer, *Complete Works* ed. W. W. Skeat III (1894) xl, 345 f., cf. II li.

[3] Boccaccio's *Genealogie* was reprinted and more or less commented on by J. Micyllus 1532. L. G. Gyraldus, *De deis gentium varia et multiplex historia* (1548), unfortunately had less effect than Conti's *Mythologia*; on both see Gruppe pp. 32 ff. and K. Borinski, 'Die Antike in Poetik und Kunsttheorie' II, *Das Erbe der Alten* 10 (1924) 29 f. Seznec (above p. 20 n. 7) pp. 229 ff.

[4] *History* [I] 237 f.

[5] *Geneal.* ed. Romano (above p. 20 n. 6) I 12.19 ff. 'Demogorgonem . . . quem profecto ego deorum gentilium omnium patrem principiumque existimo'; p. 14.27 reference to Lactantius Placidus.

[6] On Demogorgon in general and in many details see the excellent monograph of C. Landi, *Demogòrgone, con saggio di nuova edizione delle 'Genealogie Deorum Gentilium' del Boccaccio e silloge dei frammenti di Teodonzio* (Palermo 1930); on Italian poetry see pp. 7 ff. On French literature see M. Castelain, *Bulletin de l'Association Guillaume Budé* 36 (1932) 28 ff. Don Cameron Allen, *Mysterious Meant. The rediscovery of pagan symbolism and allegorical interpretation in the Renaissance* (1970) pp. 216 f., 223, 230.

[7] References are given by Castelain, loc. cit. (n. 6), in the *Oxford English Dictionary* III (1933) s.v. Demogorgon and by Seznec, op. cit. (p. 20 n. 7) p. 312 to Spenser, Robert Greene, Marlowe, Dryden, Milton, Shelley. I can add a later and less solemn one in George Meredith's early novel *Evan Harrington* (first published 1859/60), Mickleham edition (1922) p. 26 'tailordom, or Demogorgon, as the Countess was pleased to call it'.

[8] *Paradise lost* II 965; cf. 'Prolusiones oratoriae', *Opera Latina* (1698) p. 340.

god Kumarbi who may have some relation to Hesiod's Kronos,[1] but a ghostword which owes its existence to a slip of the pen. In Boccaccio's source, the scholia on Stat. *Theb.* IV 516 (the so-called Lactantius Placidus),[2] a medieval scribe corrupted 'demiurgon' to 'demogorgon'.[3] We cannot always verify the origin of strange and unique names and references in Boccaccio, as we can in this case; but there is no reason to suspect him of fictions and forgeries, since it is likely that he was able to use mythological sources lost to us.[4]

Following the example of Petrarch's *De viris illustribus* Boccaccio produced two biographical collections, *De mulieribus claris* and *De casibus virorum illustrium*,[5] in which he did not aim so much at historical truth[6] as at entertainment, mixing dry catalogues of women with spicy stories, and illustrating the tragic falls of famous men from Adam and Eve to his own time with excursuses and moral reflections. It was a stroke of good fortune that about hundred years later the great French painter Jean Fouquet and his pupils illuminated a copy of the French translation of *De casibus* with twenty-two magnificent miniatures.[7] A more modest compilation, but useful and popular for one or two centuries, was the alphabetic geographical dictionary *De montium, sylvarum, fontium nominibus*, based on Vibius Sequester; it was there that he expressed his naïve trust in the infallibility of the ancient authors:[8] when he saw with his own eyes in Italy that some of their indications were wrong, he noted: 'mallem potius eorum autoritati quam oculis credere meis.'

---

[1] *History* [I] 22.4.

[2] P. Wessner, *RE* XII (1925) 356 ff., 358.61 on Boccaccio. Cf. above, p. 21 n. 5: I cannot decide whether Boccaccio used Lactantius himself, whom he quotes, or derived his knowledge indirectly from Theodontius (see n. 4). The corrupted name is also preserved in the late medieval collection of Schol. Lucan. ed. C. F. Weber (1831) pp. 497 f.

[3] The first to restore the text was L. G. Gyraldus, *De deis gentium . . . historia* (1548) in his 'epistola nuncupatoria' pp. 2 ff. This remained unnoticed, but others (Th. Gale, C. G. Heyne) independently made the same conjecture, which is now confirmed by the reading of the best manuscript of the Stat. Schol. 'demoirgon', see Jahnke's edition (1898) and F. Cumont, *RE* V (1905) 1. The sceptics who did not accept the emendation were wrong (so Lobeck, *Aglaophamus* I [1829] 600 n., the *Oxford English Dictionary*, and G. Highet, *The Classical Tradition* (1949) p. 678, in his misleading note 51).

[4] See especially Landi, *Demogòrgone* (above, p. 21 n. 6), p. 23 on Theodontius, and my note on Call. fr. [818]; but one cannot repose the same confidence in Natalis Comes, see on Call. fr. 378 and now Jacoby, *FGrHist* III, Supplement I (1954) 240 f.

[5] On manuscripts and editions of all the minor Latin works see *Geschichte der Textüberlieferung* II 522 f.

[6] See above, p. 7.

[7] The original is one of the glories of the Bavarian State Library (Cod. gall. 6); a facsimile with translation and annotation was published by W. Pleister in 1965.

[8] See *History* [I] 232.3; cf. ibid. p. 32 on the tyranny of the book.

In contrast to Petrarch, Boccaccio stood firm to the pre-critical tradition. When he was in love with Livy,[1] he tried to translate the third and fourth Decades, not to restore the text; and when he continued the search for manuscripts of lost Latin writers, he was satisfied with the recovery of the codices without attempting any feats of textual criticism. The Tacitus of Monte Cassino (*Ann.* xi–xvi and *Hist.* i–v) is generally regarded as the most spectacular of his later discoveries;[2] but no unequivocal evidence for this has been found. No doubt Boccaccio possessed a copy of those parts of the *Annals* and *Histories* not known before the fourteenth century,[3] as he used them in making additions to his book on *Famous Women* and in the commentary on Dante written towards the end of his life; but neither he himself, though rather communicative in his letters and writings, nor any other reliable contemporary source claimed that he was the finder. The belief depends on a combination of two pieces of evidence: the romanticized story, told by Boccaccio's pupil Benvenuto Ramboldi da Imola in his immensely learned commentary on the *Commedia* of Dante,[4] of how Boccaccio visited the decaying library of the monastery Monte Cassino and burst into tears when he looked at the neglect of the precious codices; and the fact that a codex of Tacitus in Lombardic script of the eleventh century from Monte Cassino[5] was in the hands of Niccolò Niccoli before 1427.[6] It is only a guess that Boccaccio had taken away *this* manuscript from Monte Cassino and remained silent about the abstraction.[7] Equally unprovable is the assumption that he carried off from Monte Cassino the archetype of our manuscripts of Varro's

---

[1] G. Billanovich, *Giornale storico di letteratura italiana* 130 (1953) 311 ff. and M. T. Casella, 'Nuovi appunti attorno al Boccaccio traduttore di Livio', *IMU* 4 (1961) 77–129; cf. *Geschichte der Textüberlieferung* ii 520 f.

[2] R. Sabbadini, *Le scoperte dei codici Greci e Latini* i (1905) 29 f., ii (1914) 254; Tacit. ed. Koestermann i² (1965) vi f.

[3] K. J. Heilig, *Wiener Studien* 53 (1935) 95 ff. established the probability that Paulinus Venetus (d. 1344 as bishop of Pozzuoli) excerpted *Ann.* xiii–xv for his *Mappa mundi* from cod. Med. ii; he is sceptical with regard to Boccaccio.

[4] 'illud quod narrabat mihi iocose venerabilis praeceptor meus Boccaccius de Certaldo'. *Comentum super Dantis Comediam* ed. J. Ph. Lacaita v (1887) 301 f.; he heard Boccaccio lecturing on Dante in Florence in 1372, see Sabbadini (above, n. 2) ii 154.25. Cf. F. Corazzini, *Le lettere edite e inedite di Messer Giovanni Boccaccio* (1877) pp. xxxv f.

[5] Facsimile in *Codices Graeci et Latini phototypice depicti* vii 1, 2, with the preface of E. Rostagno (Leiden 1902).

[6] Poggio, *Epist.* iii 14; via S. Marco in Florence this manuscript reached the Laurentian Library, now cod. Laur. 68.2 = Mediceus II.

[7] According to Cornelia C. Coulter, 'Boccaccio and the Cassinese manuscripts of the Laurentian Library', *Class. Philology* 43 (1948) 217 ff. probably not Boccaccio, but Niccolò Acciaiuoli was responsible for the theft. On this political adventurer and collector of books see Voigt, *Wiederbelebung* i 452 ff.

De lingua Latina (Cod. Laur. 50.10);[1] but it is fairly certain that he was the first to get hold of Martial, Ausonius, Ovid's *Ibis*, parts of the *Appendix Virgiliana* and the *Priapeia*, Fulgentius, and Lactantius Placidus.[2] His commentary on the first seventeen cantos of the *Divina commedia*[3] was in large part a learned collection of biographical information about Latin authors, the first modest attempt at a modern 'history' of Roman literature.[4] Neither did he confine himself to Roman literature; in his note on the line 'Omero poeta sovrano' we find him pouring out everything he had excerpted about the 'origin, life and studies of Homer'. Boccaccio owed his little knowledge of things Greek to Leonzio Pilato.[5] He did a great practical service to classical scholarship by inviting this rather repellent man about 1360 to Florence to teach Greek; although never himself a man of means or influence in official circles, he even gave him hospitality in his own house for the three years that were devoted to the first modern translation of Homer into Latin prose.

[1] Sabbadini, *Scoperte* 1 30 f. is rather optimistic about Boccaccio's claim; on the Codex see Varro, *De lingua Latina*, ed. G. Goetz–F. Schoell (1910) pp. xiv ff.

[2] Sabbadini, op. cit. p. 33.

[3] *Il comento alla Divina commedia*, ed. D. Guerri, Scrittori d'Italia 84–6 (1918); on Homer see II 24 ff., on Horace and other Latin poets II 29 f. See also the strong criticism of Guerri's edition and the promise of a new edition by G. Padoan, *L'ultima opera di G. Boccaccio*, '*Le Esposizioni sopra il Dante*', Publicazioni della Facoltà di Lettere e Filosofia, Università di Padova 34 (1959).

[4] In the next generation Sicco Polentonus compiled the comprehensive 'Scriptorum Illustrium Latinae Linguae Libri XVIII', ed. B. L. Ullman, *Papers and Monographs of the American Academy in Rome* 6 (1928), but his *magnum opus* seems not to be indebted to Boccaccio's earlier attempt.

[5] See above, p. 14 n. 6.

II

THE SECOND AND THIRD GENERATIONS: SALUTATI, BRUNI, NICCOLI, POGGIO

It was in Florence that the first meeting of Boccaccio with Petrarch took place in 1350, that the translator of Homer became Boccaccio's guest about two years later, and that Boccaccio was chosen to give the first of the lectures on Dante founded in 1373. It was to the Florentine monastery of Santo Spirito that he finally left more than a hundred manuscripts. Petrarch and Boccaccio were men of letters, without any official status in society or in politics. They were itinerant littérateurs and had to rely on enlightened patrons. But in the next generation when the city state of Florence was firmly established, Petrarch's followers rose to the highest social and political positions.

Coluccio Salutati (1331–1406), born near Lucca and educated in Bologna in the school of rhetoric of Petrarch's friend Pietro da Muglio, became Chancellor of Florence in 1375, just after Petrarch and Boccaccio had died.[1] He had had frequent contacts with both of them, but had never met Petrarch; theirs was an epistolary friendship. In contrast to Petrarch with his more literary and abstract Roman nationalism Salutati was a real Florentine patriot and a practical politician for thirty years. The title of his opusculum *De vita associabili et operativa*[2] (which he broke off, when his wife suddenly died, and never finished) is characteristic of him, and sounds as if it was a rejoinder to Petrarch's *De vita solitaria*.[3] Yet it was Salutati's binding of the 'studia humanitatis'—a new expression[4] he liked to use—with the 'vita activa' in the service of the new city state that helped decisively to form the spirit of the Florentine Quattrocento; the best evidence for this is in the hundreds of his letters still preserved. As chancellor[5] Salutati

---

[1] B. L. Ullman, 'The Humanism of Coluccio Salutati', *Medioevo e Umanèsimo* 4 (1963); Salutati, *Epistolario*, ed. F. Novati, 4 vols. 1891–1911. See also below, p. 26 n. 8.

[2] *Epistolario* I 156 (letter to Boccaccio at the end of 1371). The identification in *Prosatori* p. 3 of this political treatise (quoted on p. 156 of Novati's edition of the letters) with the *Bucolicon carmen* (quoted ibid. p. 157) which Salutati sent to Boccaccio is a curious slip.

[3] See above, p. 15.   [4] See above, p. 16, n. 4.

[5] E. Garin, *La cultura filosofica del Rinascimento Italiano* (1961) pp. 3 ff. 'I cancellieri umanisti della Repubblica Fiorentina da Coluccio Salutati a Bartolomeo Scala'.

was the official letter writer and his letters written on behalf of the city of Florence ought to be taken into account together with his private letters (344 in Novati's edition), but there has so far been no complete collection of his public letters[1]. A famous remark is reported to have been made during the war between Florence and Milan, which started in 1390, by the Duke Gran Galeazzo Visconti that a thousand Florentine horsemen did less damage to him than the letters of Salutati.[2]

This devoted letter writer was not the discoverer of Cicero's so-called *Epistulae ad familiares*, as has been claimed,[3] but he was the first to get hold of all the sixteen books, which were copied for him in 1392 (Laur. 49.7) from a Vercelli manuscript (Laur. 49.9).[4] The influence of the complete Ciceronian corpus on humanistic epistolography can hardly be overestimated. Unlike Petrarch,[5] Salutati was able to honour Cicero's civic spirit and his taking part in the struggle for liberty instead of retreating into literary and philosophical privacy.

Salutati did not travel in search of manuscripts himself, but with the help of his numerous friends and pupils who did he was able from 1355 onwards to build up a big private library estimated to run to more than 800 volumes; 111 extant volumes have been identified.[6] He was a voracious reader and acquired a considerable knowledge of ancient Latin authors. His own letters and books show that he was also a thoughtful reader, giving his attention to problems of textual criticism[7] as well as of religion and philosophy.[8] He was not what one could call an productive scholar. But through personal contact he made the treasures of his library and his own learning accessible to others; he always frequented the meetings of the Florentine intelligentsia at S. Spirito and in the Paradiso degli Alberti, and in discussion groups he became well loved as teacher of the younger generation.

Like other young men he started his literary career with poems, of which only lines and titles quoted by himself in his letters are known to

---

[1] See Ullman, op. cit. p. 19. [2] Ibid. p. 14.

[3] This traditional mistake is repeated even by E. F. Jacob, *Italian Renaissance Studies* (1960) p. 30.

[4] Sabbadini II 214; Ullman p. 146, n. 14. Parts of the letters may have been known earlier, see above, pp. 9 f.

[5] See above, p. 10.

[6] Ullman pp. 129–209; cf. pp. 263–80 Salutati's books and their scribes.

[7] Ullman pp. 97 ff., esp. pp. 100 f. on the passage in *De fato* II 6.

[8] The philosophical elements in Salutati's study of the *litterae*, opening the way of *virtus*, are minutely examined by E. Kessler, 'Das Problem des frühen Humanismus. Seine philosophische Bedeutung bei Coluccio Salutati', *Humanistische Bibliothek*, herausgegeben von E. Grassi, Reihe 1, Abhandlungen Bd. 1 (1968). My chapter on Salutati was in final form, before the author kindly acquainted me with his new monograph. In this respect also Salutati seems to have developed a principle of Petrarch, cf. above, p. 12.

us. Although these few relics do not sound very promising, he yet maintained all his life that poetry was essentially superior to oratory. In his prose books he was not a writer of genius either, so that they did not have a wide circulation, some of them indeed having been first printed only in modern times.[1] His major work was an allegorical interpretation of the labours of Heracles, which was sketched out between 1378 and 1383 and later on enlarged into four books, but never completed.[2] The mythological parts owed much to Boccaccio, whom he greatly admired; but it seems to have been in homage to Petrarch's genius that he inserted long passionate discussions on poetry, which he ranked above all the other arts: 'poesim . . . quod merito super alias singulari promineat dignitate.'[3] This point, characteristic of the age of humanism, is particularly stressed by Politian. Salutati might almost be regarded as the first Renaissance writer on poetic theory and literary criticism.[4]

No Greek manuscript has been detected in Salutati's library,[5] and he had only the most elementary knowledge of the Greek script and some of its words; there is no evidence that he had been taught by Leonzio Pilato. But in the slow progress of Greek studies in Italy one further step was due to Salutati's energy and vigilance; he induced not a Basilian monk from south Italy, but a Greek scholar from Constantinople, Manuel Chrysoloras,[6] to accept an invitation to Florence and to teach the Greek language there from 1396 to 1400. Salutati's pupils became enthusiatic students of Chrysoloras. Perhaps the most gifted among them was *Leonardo Bruni* from Arezzo (1370?–1444), in 1427 appointed 'cancelliere dei Signori', like Salutati.[7] He was the first to make translations from Greek into Latin on a grand scale.

[1] Ullman pp. 19 ff.; cf. 'Prosatori latini del Quattrocento', *La Letteratura Italiana*, Storia e Testi 13 (1952) pp. 5 f. bibliography; only the central part of the 'Invectiva in Antonium Luschum' is reprinted pp. 7–37; cf. p. 1127.

[2] C. Salutati, *De laboribus Herculis* ed. B. L. Ullman (1951); both editions are printed.

[3] *De laboribus Herculis* p. 19.32.

[4] *A History of Literary Criticism in the Italian Renaissance* was published by B. Weinberg, 2 vols., 1961.

[5] Ullman pp. 118 ff.

[6] G. Cammelli, *I dotti Bizantini e le origini dell' Umanesimo* i: *Manuele Crisolora* (1941) pp. 28 ff.

[7] E. Garin in 'Prosatori latini del Quattrocento', *La Letteratura Italiana*, Storia e Testi 13 (1952) 39 ff. with bibliography and with a part of the text of the *Dialogi* (see below, p. 30); *Epistolarum libri VIII* ed. L. Mehus, 1741; L. Bruni Aretino, 'Humanistisch-philosophische Schriften' mit einer Chronologie seiner Werke und Briefe. hrsg. und erläutert von H. Baron, *Quellen zur Geistesgeschichte des Mittelalters und der Renaissance* i (1928). The collection of material is useful, but the deficiencies of the editing and explanation of the text are lamentable, see L. Bertalot, *Archivium Romanicum* 15 (1931) 284–323; on Baron's later publications see Garin, loc. cit. and W. K. Ferguson, *Journal of the History of Ideas* 19 (1958) 14–25 and Baron's answer ibid. pp. 26–34.

Since the censor Appius Claudius Caecus at the beginning of the third century B.C. first translated Greek maxims into Latin, translations of Greek literature had been of vital importance for Roman culture. The Romans were the first translators in Europe, they became the translators κατ' ἐξοχήν. Translations from Greek were indispensable in early times when knowledge of the Greek language was not yet firmly established, and again when after the bilingual centuries of the Roman empire Greek was slowly disappearing from the West in the fourth and fifth centuries A.D.[1] It was a sort of patriotic duty of certain aristocratic circles to prevent it from dying out completely. The noble politician and philosopher Boethius conceived a gigantic plan for translating the whole of Aristotle and even of Plato, which was frustrated by his execution in 524. The later medieval Latin translations, especially those of the twelfth century, sounded peculiar to the ears of Petrarch and the other humanists used to the *sonoritas* of the Roman classics, and they therefore tried all the harder to achieve an authentic style. The misfortune was that *Latinitate donare* sometimes meant Romanizing the poor Greek writers and embellishing them with an un-Greek rhetorical style. But no doubt such transformations struck the Italian mind much more than the original would ever have done.

Bruni was prudent and modest enough to confine himself almost exclusively to translations of Greek prose. One exception was the three speeches in *Iliad* IX 222–603 which he translated 'oratorio more' into Latin prose, with a poem[2] on Homer whom he characterized as 'paene perfectus in eloquentia'. As a scholar knowing Greek, Bruni could hardly have been unacquainted with Homer;[3] but an unexpected second exception was a rendering of Aristophanes *Plutus* 1–269 in Latin prose.[4] This latest piece of classical Attic poetry, moralistic and allegorical, had always been preferred from early Hellenistic times to all the other plays,[5] and was now the first to be made known to the Western world. We cannot date Bruni's Latin *Plutus*. But it may have had some connection with a Latin paraphrase of another part of the same play. In 1416 or 1417 Rinucci (who later taught the elderly

---

[1] Marrou 262; cf. P. Courcelle, *Les Lettres grecs en occident de Macrobe à Cassiodore*, 2nd ed. 1948.

[2] Baron (1928) pp. 132–4; cf. Pertusi, *Leonzio Pilato* p. 532.

[3] On Barlaam and Leonzio Pilato see above, pp. 14 ff.

[4] D. P. Lockwood published the text in *Classical Studies in Honor of J. C. Rolfe* (1931) pp. 163 ff.; Creizenach in Koch's *Studien zur vergleichenden Literaturgeschichte* IV (1904) 385 f. had already referred to Bruni's translation, see also W. Süß, 'Aristophanes und die Nachwelt', *Das Erbe der Alten* 2/3 (1911) 23. Baron (above p. 27 n. 7) unfortunately overlooked it.

[5] *History* [1] 161.

Poggio and the young Valla some Greek) told his friends in Crete a story which contained a paraphrase of *Plutus* 400–626 'Penias fabula'; and one of his listeners, Cristoforo de' Buondelmonti, preserved it in his 'Descriptio Candiae',[1] the earliest book of archaeological travels, which is dedicated to Buondelmonti's teacher, Bruni's great friend Niccolò Niccoli.

It is natural that a translator of that age should soon have turned to Plutarch's works,[2] which had been lost to the medieval Western world; for the *Lives* of Plutarch strongly appealed to the feeling of the Italian Renaissance for the individual, in particular those of the great Romans; and to a lesser degree his *Moralia* appealed to its concern for problems of moral philosophy. We should not forget Salutati's uncommon interest in Plutarch, often expressed in his correspondence with Chrysoloras,[3] and his stimulating influence on the younger generation though now giving credit to Bruni.[4] Bruni's Latin Plutarch was the basis of the translations into the vernacular, and so it is thanks to him that Plutarch's writings became known and remained popular for centuries. Bruni went on to Xenophon 'praecipuo quodam amore', to the orators, and to the philosophers. Between 1405 and 1435 he translated not only six dialogues of Plato, but also some of his *Letters*; in his dedication of them to Cosimo de' Medici he expressed[5] his enjoyment of the feeling of personal contact with the ancient author. His most discussed translations were those of Aristotle's *Politics* and *Ethics*; in his introductions as well as in a separate treatise *De interpretatione recta*[6] he vehemently attacked the medieval translations and tried to justify his own method,[7] based on the intimate comparative study of both languages.

[1] Flam. Cornaro, *Creta sacra* (1755) I 94; see D. P. Lockwood, *Harvard Studies in Classica Philology* 24 (1913) 52, 72 ff. and E. Jacobs, 'Zu Buondelmontis kretischen Reisen', *Stephaniskos für E. Fabricius* (1927) p. 60.

[2] A precise chronological arrangement is not yet possible in spite of the work of Baron and Garin.

[3] Novati, *Epistolario* IV 336.1 and 682 (Index); cf. D. P. Lockwood, 'Plutarch in the 14th century', *TAPA* 64 (1933) lxvi f.

[4] R. Hirzel, 'Plutarch', *Das Erbe der Alten* IV (1912) 102 ff., esp. 106 f.; K. Ziegler, *RE* XXI (1951) 953. See also G. Resta, 'Le epitomi di Plutarco nel quattrocento', *Miscellanea erudita* V (1962): on humanistic epitomes (esp. Pier Candido Decembrio), the enormous popularity of the *Lives*, and Salutati's proper merits.

[5] E. Garin, 'Medioevo e Rinascimento', *Biblioteca di cultura moderna* 506 (1954) 122.21 'traductio . . . ita vehementer mihi iocunda fuit', etc.

[6] Baron, op. cit. pp. 70 ff. reprinted the introductions and the treatise.

[7] One is tempted to compare St. Jerome's *Epist.* 106 in which he speaks of his principles of translating, *CSEL* 55 (1912, reprinted 1961) 250 'hanc esse regulam boni interpretis, ut ἰδιώματα linguae alterius suae linguae exprimat proprietate'.

It was Bruni who rediscovered the ancient prose rhythm and discussed at length Cicero's theory and practice; previous generations had ignored the 'numerosa structura' despite its importance for the humanistic imitation of Ciceronian Latin.[1] He also brought to light Aristotle's references to rhythm in prose.

Long before he held the chancellorship of Florence, Bruni began in 1404 to write his *Historiarum Florentini populi libri* in twelve books of which the last remained unfinished at his death in 1444.[2] This great work has been aptly termed 'Humanistic Historiography'; it is strongly influenced by the new scholarship of Petrarch and Salutati and shows a wide knowledge not only of Latin, but also of Greek literature, and it passionately defends the principle of political liberty. An attractive picture of his learned Florentine friends is given in his *Dialogi ad Petrum Paulum Histrum* in which he revived the style of Cicero's dialogues; we have already referred[3] to the impressive passage where Niccolò Niccoli praised Petrarch as the restorer of 'studia humanitatis'.

Niccolò Niccoli (1363–1437) was one of the most distinguished members of this Florentine circle. He never became a public figure like Salutati and Bruni, but preferred a retired life; an aesthete of the most discriminating taste, he enjoyed the beauty of all ancient things in tranquillity. His sense of style was so sublime that he hardly dared to write or speak in Latin himself; his letters were written in his native tongue, and so was the only scholarly treatise he produced, on Latin orthography. Indeed the inevitable consequence of such perfectionism is that Latin becomes a dead language. Niccoli was indefatigable in collecting and copying manuscripts with his own hand;[4] but he also collated them, compared the text of different manuscripts, and arranged it in paragraphs, with the addition of headlines, a work, it must be said, of bibliophily rather than textual criticism.[5] Niccoli was not a great traveller, but he gave valuable instructions to travelling friends and agents of the Medici. One of his lists is by chance preserved,[6] an exciting

---

[1] Bruni, 'De studiis et litteris liber', Baron, op. cit. p. 10 'Omnis . . . oratio pedibus suis commovenda erit; quos si ignoret scribens, velut in tenebris ambulet necesse est'; cf. Zielinski, *Cicero im Wandel der Jahrhunderte*[2] (1908) p. 424. It is a slip in Sandys, *Harvard Lectures* (1905) p. 158, to say 'that Cortesi had discovered the importance of a rhythmical structure in the composition of Ciceronian prose', *De hominibus doctis* ed. Galletti p. 23.

[2] Critical text in *Rerum Italicarum Scriptores* XIX 3 (1914–26) ed. E. Santini e C. di Pierro. B. L. Ullman, 'L. Bruni and humanistic historiography' in *Studies in the Italian Renaissance = Storia e Letteratura* 51 (1955) pp. 321–44.

[3] See above p. 16, n. 3.   [4] On humanistic script see above, p. 13.

[5] His textual 'criticism' seems to be overestimated by H. Rüdiger, *Textüberlieferung* I 552.

[6] *Commentarius Nicolai Niccoli*; a first hint to its existence was given by E. Jacobs, *Wochenschrift für klassische Philologie* (1913), p. 701; the original is now in the Pierpont Morgan

document in that it contains the minor works of Tacitus which were still missing in Niccoli's time. They had been listed in the 'Inventarium monachi Hersfeldensis' sent by this monk to Niccoli, and Niccoli dispatched an envoy to Germany to fetch the Tacitus codex from the great library of Fulda.[1] It should really be called the Codex Fuldensis, not Hersfeldensis as it usually is. Its adventurous story cannot be retold here; for our purpose what matters is the initiative taken by Niccoli in order to get classical manuscripts from abroad.

The mastery in this field was soon won by *Poggio Bracciolini* (1380–1459), not only the most active and the luckiest hunter of manuscripts in Western Europe, but at the same time a great letter-writer and teller of short stories, one of the liveliest figures of the age.[2] The office of papal secretary, which Poggio occupied much longer than Salutati and Bruni, was of great consequence to him; for in this capacity he had to attend the Council of Constance 1414–18, and when the Holy See became vacant for two and a half years he was free to make four long expeditions. From Constance, situated in the centre of Western Europe, his second and third journeys led him to the library of the monastery at St. Gallen only fifteen miles away and perhaps to Einsiedeln, the first and fourth journeys to French and German monasteries and cathedrals. He succeeded in detecting a tremendous number of lost Latin authors and in bearing them off as prizes, or at least in having the text copied. Two

Library, New York and its text is published by R. P. Robinson, 'De fragmenti Suetoniani de grammaticis et rhetoribus codicum nexu et fide', *University of Illinois Studies in Language and Literature* VI 4 (1922).

[1] This was proved by L. Pralle, 'Die Wiederentdeckung des Tacitus', *Quellen und Abhandlungen zur Geschichte der Abtei und der Diözese Fulda* 17 (1952) 15 ff. 'Heinrich von Grebenstein [the "monachus Hersfeldensis"] und die Entdeckung der kleinen Tacitus-Schriften', and especially pp. 42 ff. on the codex brought from Fulda to Italy. Other parts of the book, to which E. Arnold referred me, are unfortunately disfigured by an awkward Panfuldaism and a strange lack of knowledge and criticism. My impression that he was right in this one particular point was confirmed by B. Bischoff.

[2] E. Walser, *Poggius Florentinus, Leben und Werke* (1914). The biography (pp. 1–324) is followed by 'Dokumente' and 'Inedita' (pp. 325–560); the inventory of his books at the time of his death is 'a document of capital importance for the transmission of classical texts', as A. C. Clark said in 'The Reappearance of the Texts of the Classics', *The Library* 4th ser. 2 (1921) 36. An important supplement is the list of manuscripts copied by Poggio either in his formal bookhand (10 items) or in his less formal hand (13 items), see B. L. Ullman, *Studies in the Italian Renaissance* (1955) pp. 315 ff. Walser prepared a new critical edition of Poggio's letters (see above p. 18 n. 5); Helene Harth, a pupil of W. Rüegg, is now expected to finish this important great enterprise (see *Prosatori*, p. 1129). Meanwhile we have available Poggii *Epistolae* ed. Th. de Tonellis, 3 vols., 1832–61, reprinted 1964 as Tomus III of Poggio's 'Opera omnia' = *Monumenta politica et philosophica rariora*, Ser. II, no. 6; see ibid. vol. I, pp. xiii ff. the 'premessa' of R. Fubini on the 'epistolario'. The reprint of 'Opera omnia' just quoted reproduces the Basle edition of 1538. A selection of letters and writings in *Prosatori* ed. Garin (1951) pp. 215 ff. Cf. R. Sabbadini, *Scoperte* I 77 ff.

generations after Boccaccio[1] it seems to have become almost a common-place that the Italian humanist, beholding the filthy state of neglected codices, should weep and regard it as his duty to free them from their prison. Sabbadini[2] correctly distinguished Poggio's four travels, and in spite of A. C. Clark's scepticism with regard to Poggio's first expedition to the Abbey of Cluny, near Mâcon in Burgundy, there is no wholly convincing argument for rejecting his own unequivocal statement: 'orationes Tullii . . . quas detuli ex monasterio Cluniacensi',[3] that is Cicero's five speeches *pro Milone*, *pro Cluentio*, *pro Murena*, *pro Sexto Roscio*, and *pro Caelio*. This manuscript, called the 'Vetus Cluniacensis',[4] was the first sent by Poggio to Niccoli, and must have reached Italy before the end of 1415; the two new speeches, *pro S. Roscio* and *pro Murena*, caused a sensation, and the text of the other three was a considerable improvement on that previously known. The result of his fourth journey was the discovery of no fewer than eight Ciceronian speeches, one in Langres, *pro Caecina*, and the others in the 'small' library of Cologne cathedral. The original of the Cluniacensis was later lost, and its readings have had to be reconstructed from copies made by various hands; of his second great Ciceronian find Poggio had made a transcript by his own hand, which also disappeared, but was quite recently rediscovered by A. Campana in the Vatican library.[5]

This autograph is an excellent specimen of his beautiful handwriting, which played a decisive part in fixing the new style of the 'littera antiqua'.[6] Poggio generally tried to produce a readable text and to correct the obvious scribal errors; though he proudly called this activity 'emendare' in his letters, he hardly achieved more in this respect than Salutati and Niccoli.[7] From his two trips to St. Gallen in 1416 he returned with various new treasures: Asconius' commentary on Cicero, of exceptional interest to the Italian Ciceronians, the first complete manuscript of Quintilian, and part of Valerius Flaccus' *Argonautica*; from his fourth expedition to France and Germany in 1417 he brought

[1] See above, p. 23.
[2] *Scoperte* 1 77; Walser, op. cit. pp. 49 ff. agreed with Sabbadini; A. C. Clark, 'The Reappearance of the Texts of the Classics', *The Library*, 4th ser. 2 (1921) 26 f. did not.
[3] *Epist.* 1 100, cf. 153.
[4] A. C. Clark, 'The Vetus Cluniacensis of Poggio', *Anecdota Oxoniensia*, Class. Ser. 10 (1905).
[5] Cic. *In Pisonem* ed. R. G. M. Nisbet, Oxford 1961, p. xxv, Cod. Vatican. Lat. 11458; Campana 'himself intends to give a full account in due course'.
[6] See above, p. 13 with n. 8.
[7] See above, pp. 26 f. H. W. Garrod, *Scholarship* (1946) p. 23 went too far in his admiration of Poggio, when he regarded him as 'the founder of modern scholarship' and praised his 'fine critical discernment'; for a balanced judgement see H. Rüdiger, *Textüberlieferung* 1 553.

manuscripts of Lucretius,[1] Silius Italicus, Manilius, and Statius' *Silvae*. But Cicero was ranked incomparably higher than all the others. In 1421, shortly after Poggio's discovery of the series of new speeches, the bishop of Lodi, Gerardus Landriani, found the triad of rhetorical writings:[2] *De oratore* complete, *Orator*, and *Brutus*. We owe about half of all Cicero's writings that we now possess to Petrarch and Poggio; and these precious new books were not just a welcome enlargement of literary knowledge, but a quickening force of life.

When the Council of Constance broke off in May 1418, Poggio, after some hesitation, followed Cardinal Henry Beaufort, bishop of Winchester, to England, where he lived for four years in different places, but never in Oxford. So he felt like an exile, deeply depressed by the foggy sky and uncivilized people, 'quorum Deus venter est', as he said.[3] No classical manuscripts could be found except a 'particula' of Petronius,[4] and in his melancholy Poggio turned to the study of the Church Fathers,[5] with profit to his later treatises on moral philosophy but not to classical scholarship. When he returned to the Vatican chancery, he was again in high spirits, and began to include archaeology among his interests.[6] In 1453 he settled down happily in his own city of Florence as chancellor like Salutati and Bruni.[7] And like Bruni he concentrated himself, between his official duties, on writing his *Historia Florentina* of the last hundred years.

But all his life he had been a writer of distinction in Latin prose, and his anecdotes in the *Liber facetiarum* became extremely popular. His mastery of Latin is equally clear in his letters, whether in the obvious pleasure with which he described to Niccoli the gay social life of the Swiss bathing resort Baden or in the deep emotion of the vivid account of the trial and execution of Jerome of Prague in a letter to Bruni, full of classical echoes.[8] Poggio published a number of treatises, mostly in dialogue form, and funeral speeches on Niccoli, Bruni, and others.

---

[1] On Lucretius see Konrad Müller, 'De codicum Lucretii Italicorum origine', *Museum Helveticum* 30 (1973) 166–78.

[2] J. Stroux, *Handschriftliche Studien zu Cicero De oratore* (1921); on the date p. 8; the codex disappeared from sight in 1428 and has to be reconstructed from descendants.

[3] *Epist.* I 64.

[4] On his return journey to Italy he found another codex in Cologne; on both manuscripts with excerpts see Konrad Müller in his critical edition of Petronius (1961) p. viii f.

[5] Walser, *Poggius* pp. 79 ff.   [6] Cf. below, p. 50.

[7] N. Rubinstein, 'Poggio Bracciolini, cancelliere e storico di Firenze', *Atti e Memorie della Academia Petrarca* N.S. 37 (1965) 215 ff.; cf. above, p. 25 n. 5.

[8] *Prosatori* p. 238 'stabat impavidus, intrepidus, mortem non contemnens solum, sed appetens, ut alterum Catonem dixisses. O virum dignum memoria hominum sempiterna. . . . nullus unquam Stoicorum fuit tam constanti animo, tam forti mortem perpessus.'

When he had learnt some Greek, mainly from Rinucci,[1] he followed Bruni's example by trying his hand as translator; by August 1449 he had completed a Latin translation of the first five books of Diodorus, commissioned by Pope Nicolas V.[2] Being a rather quarrelsome man, he was involved in some literary feuds and occasionally wrote furious invectives. Poggio never strove after a Ciceronian style or even a grammatically correct Latin; he treated Latin as if it were a living language, and because of that we see him in the last years of his life at feud with the leading spirit of the next generation, Lorenzo Valla.

One day in 1451 Poggio found in a copy of the collection of his letters to Niccoli, of which he was very proud, some critical and ironical comments upon his Latinity, scrawled in the margin by a pupil of Valla's; he got so angry with Valla, whom he suspected of being the author, that he tried to have him murdered,[3] a dramatic refutation, had he succeeded, of Schopenhauer's saying that 'the history of . . . learning and art' (in contrast to the universal history of the world) 'is always going on . . . guiltless and without bloodshed.'[4] But Poggio finally confined himself to a form of retaliation more appropriate in a scholar, a literary invective. Valla, no less pugnacious, replied, and a war of pamphlets, five from each side, ensued, the arguments of which were of a general importance far beyond the trivial cause.

[1] See above, p. 28.

[2] The *Bibliotheca historica* of Diodorus Siculus, translated by John Skelton, now first edited by F. M. Salter and H. L. R. Edwards, ii (1957) xx ff. 'Poggio's translation'.

[3] Walser, *Poggius* pp. 273 ff.

[4] *A. Schopenhauers Sämtliche Werke*, hrsg. v. A. Hübscher [vol. 6:] 'Parerga und Paralipomena' 2. Bd. (1947) p. 79.

III

LORENZO VALLA

Lorenzo Valla (1407–57)[1] was probably born in Rome, where after wanderings through many Italian cities and a period in the service of king Alfonso of Aragon and Sicily at Gaëta and Naples from 1435 to 1448 he finally settled as papal scriptor and secretary until his death. At the centre of his scholarly work were the long and intensive studies of the Latin language completed about 1440 under the title *Elegantiae Latini sermonis*; 59 editions were printed between 1471 and 1536. The preface contains the highest praise of the Latin language[2] ever written. This language is eternal like Rome itself; indeed, the empire is lost, while the language still lives. But even the language has deteriorated much since the Gothic invasions of Italy. (It was thanks to Valla that 'Gothic', even in a stylistic context, became a term of abuse.)[3] The Latin language has to be restored in order to restore the old splendour of Rome. This was in the line of Petrarch's nationalism; the new idea, however, was to fix a strict definition of the ancient 'usus loquendi' from Cicero and Quintilian, and to demand that the recognized *veritas*, the truth, should be valid for the present and for all time.[4] Since Poggio would never have dreamed of accommodating his beloved Latin to such demands, their quarrel was hardly avoidable. The most brilliant piece

---

[1] L. Valla, *Opera omnia*, con una premessa di E. Garin (Torino 1962) 1: Scripta in ed. Basilensi anno MDXL collecta. II: 1. De rebus a Ferdinando gestis—63. De mysterio Eucharistiae—73. Opuscula quaedam (1503)—131. Opuscula tria ed. J. Vahlen 1869: Oratio in principio sui studii—De professione religiosorum—Praefatio in Demosthenem—339. Encomium S. Thomae—353. Epistolae et documenta—465. Oratio ad Alphonsum regem—475. Epistulae. Particularly important is the reprint of R. Sabbadini, 'Cronologia documentata della vita di Lorenzo della Valle, detto il Valla', Firenze 1891 in vol. II of the reprinted *Opera*, pp. 353–454 under the heading 'Documenta'. We fervently hope for a new critical edition of Valla's *Opera*; we must accept Sabbadini's chronology, until a new biography is written on the basis of a new edition.—A small selection from the *Elegantiae, De libero arbitrio*, and *De professione religiosorum* is printed with an introduction by E. Garin in *Prosatori Latini del Quattrocento* (1952) pp. 521–631. See also the monograph of F. Gaeta, *L. Valla, Filologia e Storia nell' Umanesimo Italiano* (1955), esp. ch. III 'La nuova filologia e il suo significato'. Salvatore J. Camporeale, *Lorenzo Valla* (1972).

[2] Cf. Gorgias' hymn on the λόγος, *History* [1] 49.

[3] *Elegantiae* lib. III praef. See Erasmus, *ep.* 182.79 ff. on Valla and the Goths.

[4] 'Ego pro lege accipio quidquid magnis auctoribus placuit', taken out of G. Funaioli, *Studi di letteratura antica* I (1951) 278.

in this exchange of scurrilous invective was a so-called 'apologus' of
Valla, a dramatic scene in which the great Italian educationalist
Guarino recites passages of Poggio's Latin letters, and his cook and his
groom are to judge the Latinity: Poggio is said to have used 'culinaria
vocabula', to have learnt his Latin from a cook; as a cook breaks pots
to pieces, so he smashes grammatical Latin. No doubt the terms
'Latinum culinarium', 'Latin de cuisine', 'Küchenlatein' are derived
from this amusing humanistic invention,[1] not from lampoons against
monastery kitchens at the time of the Reformation.

There are two reasons for dwelling upon these polemics. They mark
the turning-point in the history of modern Latin; we can clearly see in
them the beginning[2] of the accurate study of that Latin language and
style which ought to be used by speakers and writers, and the approach-
ing end of Latin as a freely living uncontrolled language. And secondly,
the passionate distinction between good Latin as the truth and bad
Latin as falsehood is fundamental for Valla in nearly all his writings.
His so-called criticism always starts from questions of the Latin
language, a point that has not always been understood by modern
readers.

It is relatively easy to appreciate this in his textual criticism of Latin
authors; but in the case of his biblical criticism and historical criticism
it is surprising and harder to grasp. Thanks to recent discoveries we
can now see exactly the unique spectacle of how Valla added his own
critical notes to Petrarch's annotated manuscript of Livy.[3] This is an
exciting supplement to his *Emendationes Livianae*, in so far as it contains
the earlier notes on which the conjectures finally published in the
Emendationes were based. King Alfonso, having firmly established his rule
after years of war, had assembled poets and scholars at his court in
Naples. We remember that questions on difficulties in Homer and their
solutions had amused Ptolemaic kings in Hellenistic times.[4] Now again
an inquisitive sovereign liked to put questions on ancient texts and to
take part in the debates of his learned circle. Lorenzo Valla was the
most distinguished of its scholarly members, and Livy was a favourite
subject of the discussions; for Cosimo de' Medici had recently, at the

---

[1] R. Pfeiffer, 'Küchenlatein', *Philologus* 86 (1931) 455 ff. = *Ausgewählte Schriften* (1960)
pp. 183 ff.
[2] Cf. above p. 34.
[3] See above, p. 8 and especially n. 3; Petrarch's and Valla's handwriting on plates 30–2
of Billanovich's article. Cf. also Billanovich and others, 'Per la fortuna di Tito Livio nel
Rinascimento', *IMU* 1 (1958) 245 ff. and pl. XVI.
[4] See *History* [1] 70.

end of 1444, presented Alfonso with a beautiful manuscript of Decades
I, III, and IV, the so-called 'Codex Regius'.[1]

Always in fighting spirit, Valla liked to ridicule the vain efforts of
others towards restoring the corrupt text and then to produce his own
emendation, often in triumphant mood. It is surprising in how many
cases elation was justified. In Livy XXI about twenty emendations of
Valla have been generally accepted;[2] so has his 'reficiuntur' for the
reading 'refiguntur' of his codex in XXIII 34.17, while very bold
changes like 'scutorum' for 'suetaeque' XXIII 30.3 have been duly
rejected by modern scholars.[3] We shall be able to appreciate the
different stages of Valla's critical work on Livy's text when a new text
and a commentary on his *Emendationes* are available.[4] But clearly it is
his work, not Poggio's,[5] that marks the decisive stage in the reintroduction
of textual criticism.

In Livy Valla's polemics were directed against his contemporaries
who in his opinion depraved the text: 'vos . . . regium codicem . . .
depravatis';[6] in his work upon the Vulgate of the New Testament[7] he
was making war on the great translator of late antiquity when he
compared St. Jerome's Latin version with the Greek original and made
a register of his mistakes; the main object of his very severe criticism
was the Latin style of the Church Father. It has been assumed that
Valla entered the field of biblical criticism under the influence of the
Byzantine tradition, and was in particular dependent on the critical
treatment of the Greek and Latin text of St. John. 21:21 ff. by his friend
Bessarion of Trapezus.[8] But as Bessarion's paper is dated by its editor
not earlier than 1455 (by others much later), the chronological
difficulties are insurmountable; for Valla wrote his *Adnotationes* in the

[1] See *IMU* I (1958) 245 ff. and pl. XVI.

[2] R. Sabbadini, 'Il metodo degli umanisti', *Bibliotechina del 'Saggiatore'* 3 (1920) 59.

[3] Valla, *Op.* pp. 603–20 'Emendationes in Livium'; pp. 612 f. the two passages quoted.

[4] Mariangela Ferraris is said to be preparing the new edition of the *Emendationes*, see
G. Billanovich, *IMU* I (1958) 275.2.

[5] See above, p. 32 n. 7.   [6] *Op.* p. 612.

[7] *Op.* pp. 801–95, 'Adnotationes'.

[8] L. Mohler, 'Kardinal Bessarion als Theologe, Humanist und Staatsmann' Bd. 3, *Quellen
und Forschungen aus dem Gebiete der Geschichte*, hg. von der Görres-Gesellschaft 24 (1942, repr.
1967) 70–90 Text. Mohler says 'daß diese Schrift Lorenzo Valla die Anregung zur Abfassung
seiner *Adnotationes* zum Neuen Testament gegeben hat' (p. 70); cf. Mohler, Bd. 1 = *Quellen* 20
(1923) 403. Bessarion, the most important of the Greek immigrants, remained in Italy after
the Council of Florence (1439), joined the Roman church, and was made a cardinal. It was
he who in 1461 took Regiomontanus, the first editor of Manilius and the translator of Ptolemy,
as his companion to Rome, giving him access to his manuscripts of Greek astronomy and thus
furthering the renaissance of science. On the first printed edition of 1538 see below, p. 139
(Camerarius).

forties, probably in 1448. In any case the difference in approach between Bessarion and Valla is decisive; the cardinal was contributing to the traditional theological discussion of the passage, while the author of the *Elegantiae Latini sermonis* had adopted the novel course of examining the translator's Latin to see how far it was in harmony with the 'truth'[1] and how often he had failed.

Erasmus,[2] while 'hunting' in libraries, found a manuscript of the *Adnotationes* in the Praemonstratensian Abbey of Parc near Louvain, and was persuaded in 1505 by his British friend Christopher Fisher, at this time papal protonotary, to give it to a Paris printer. Erasmus's preface, written in a partly exuberant, partly ironical style is still by far the best essay on Valla, in whom he recognized a kindred soul. The knowledge of Valla's critical notes was of the highest value for Erasmus; for, although it was not ready for publication until 1516, he had already taken the first steps towards the discovery of the 'veritas evangelica', in preparing his greatest scholarly work, his own Latin translation and his edition of the Greek text of the New Testament.

Valla himself had considerable experience as a translator of Greek poetry and prose into Latin. In 1428/9 he had started to translate four books of the *Iliad*[3] into a clear Latin prose, and he managed to Latinize two-thirds of the Homeric poem between 1442 and 1444; the work was completed after his death by his pupil Francesco Aretino. Pope Nicolas V, who was very fond of the Greek historians, commissioned a series of scholars to translate them into Latin;[4] Valla's part was the most difficult one, Thucydides. When he had finished it in 1452 after about four years' hard work, he was called by the pope to do the same for Herodotus, and his translation was just completed when he died in 1457, but unrevised and without a proem.[5]

The assumption that his intimate contact with Thucydides aroused fresh powers of 'historical judgement' in Valla's mind and thus equipped

---

[1] On 'veritas' see above, p. 35.

[2] Erasm. *ep.* 182.1 ff. Allen: 'quum in pervetusta quapiam bibliotheca venarer (nullis enim in saltubus venatus iucundior), forte in casses meos incidit praeda neutiquam vulgaris, Laurentii Vallae in Novum Testamentum annotationes.' On the many references to Valla in Erasmus's letters see *Opus epistularum* XII (1958) 180; the earliest characteristic reference is *ep.* 29.18 f. 'Ego illius doctrinam qua meo iudicio nulla probatior, tuendam mihi sumpsi'.

[3] G. Finsler, *Homer in der Neuzeit* (1912) pp. 28 f.—On Bruni see above, p. 28.

[4] On Poggio's Diodorus see above, p. 34.

[5] Giovan Battista Alberti, 'Tucidide nella traduzione Latina di L. Valla', *Studi italiani di filologia classica* 29 (1957) 1–26; G. A. Alberti, 'Erodoto nella traduzione Latina di L. Valla', *Bollettino del Comitato per la preparazione della Edizione Nazionale dei Classici Greci e Latini*, N.S. 7 (1959) 65–84.—On translations of smaller excerpts from Aesop, Demosthenes, and Xenophon see L. Valla, *Opera omnia* ed. E. Garin (above, p. 35 n. 1) I, p.v.

him to compose his most remarkable book, the so-called *Declamatio de falso credita et ementita Constantini Donatione*, was a regrettable mistake of Wilamowitz.[1] There is first of all once again a difficulty of chronology: Valla began his work on Thucydides late in his life, certainly not before 1448, while the *Declamatio* could hardly have been written later than 1440 when he was in the service of king Alfonso. Thucydides' critical judgements are based on careful inferences from comparison, εἰκάζει, and demand τεκμήρια and σημεῖα, but Valla's main arguments against the authenticity of the *Declamatio* are based on the analysis of its Latin language, which is very bad Latin indeed. His 'method' is the same as in his other writings and is separated by a world of difference from that of Thucydides.

It is quite different even from that of a contemporary who had made an attack on the *Constitutio* some years earlier, during the Council of Basle; this was Nicholas of Cusa in book III chapter 2 of his great work *De concordantia catholica*,[2] which was finished towards the end of 1433. The possibility that Valla got to know that chapter cannot be ruled out, but there is no evidence of any direct influence although the two men were on friendly terms.[3] Nicholas examined the *Constitutio* from the point of view of church history and theology. Valla, on the other hand, discovered from a scrutiny of the manuscript tradition that the passage on the Constantine donation was missing in the oldest manuscripts of the *Decreta* and apparently interpolated in later manuscripts; he also realized that the debased Latin belongs to a much later time than that of Constantine.[4] When he came to this point, he lost patience and raged against the scoundrel: 'O scelerate atque malifice' (p. 34.26 Schwahn) . . . 'huic asino tam vaste immaniterque rudenti' (p. 37.7 Schw.) . . . 'oratio . . . alicuius clericuli stolidi, . . . saginati et crassi, ac inter crapulam interque fervorem vini has sententias et haec verba ructantis' (p. 57.9 Schw.).

---

[1] Wilamowitz, 'Antike und Hellenentum', *Reden und Vorträge* II (4th ed. 1926) 115 'Zuerst griff man nach den Historikern, und sofort zeigte sich, wie die Berührung mit Thukydides auf das geschichtliche Urteilsvermögen wirkte: sein Übersetzer Lorenzo Valla durchschaute die Fälschung der Konstantinischen Schenkung'; cf. 'Geschichte der Philologie', *Einleitung in die Altertumswissenschaft* I I (1921) 12.

[2] Nicolai de Cusa, *Opera omnia* XIV (1939) 'De concordantia catholica' ed. G. Kallen, pp. 328 ff.

[3] G. Laehr, 'Die Konstantinische Schenkung in der abendländischen Literatur des ausgehenden Mittelalters', *Quellen und Forschungen aus italienischen Archiven und Bibliotheken* 23 (1931/2) 157 ff. H. Fuhrmann, 'Zu Lorenzo Vallas Schrift über die Konstantinische Schenkung', *Studi medievali*, Ser. 3, anno 11 (1970) 913 ff. Cf. below, p. 40 n. 1.

[4] 'Illa loquendi barbaries nonne testatur non saeculo Constantini, sed posteriori, cantilenam hanc esse confictam' (p. 51.32 Schw.).

The *Declamatio* was printed by Ulrich von Hutten at the beginning
of the Reformation in 1517 as a polemic against the Papacy;[1] this is one
of Hutten's typical mistakes and has nothing to do with Valla. Valla's
aim was to remove a dangerous misunderstanding about the secular
power of the Pope and to reach the truth of his position ('edoctus
veritatem' p. 82.9 Schw.). 'Tunc papa et dicetur et erit pater sanctus
[the Holy Father], pater omnium, pater ecclesiae; nec bella inter
Christianos excitabit, sed ab aliis excitata censura apostolica et papali
maiestate sedabit' (p. 82.21 ff. Schw.). This is the solemn and peaceful
conclusion of a treatise so often satirical and aggressive; there is no
reason to be surprised that Valla remained on good terms with the
Curia and died as papal secretary.

Two other minor problems of authenticity seem to have been raised
by Valla. He expressed doubts about the general belief that the writings
under the name of Dionysius Areopagita were the work of a pupil
of St. Paul; the final proof was given by J. Scaliger.[2] Valla also published
a treatise about the famous correspondence of St. Paul and Seneca
('de ementitis . . . epistolis alio opere disputavimus'),[3] in which he
seems to have been the first to say that it could not be genuine; but
this 'aliud opus' was lost and has not yet been rediscovered.

Valla's literary accomplishments were not confined to his works of
pure scholarship. He was, as we have seen a consistent thinker by
nature, and no doubt the philosophical studies of his early years
strengthened his inborn faculty. He started in 1431 with ethical
problems in his treatise *De voluptate* (on Epicureanism and Christianity),
which he reshaped and enlarged in 1433 and later; he went on to logic
in his *Dialecticae disputationes contra Aristotelicos* (1433 to 1438); and then
he returned to questions of ethics and religion in his *De libero arbitrio* and
De professione religiosorum, written before 1442.[4] Though everything he

[1] A. C. Clark, *Cl. R.* 38 (1924) 88 found in the Bodleian Library a copy of the *Declamatio*
printed 'per Anonymum de Aloysio', 1506. But is this date correct?—Clark in this review of
C. B. Coleman, *The Treatise of Lorenzo Valla on the Donation of Constantine*, Text and transl. into
English (1922), ventured to suggest that Nicholas of Cusa 'borrowed from Valla, not *vice
versa*', which was the *communis opinio*. But it is highly improbable that the young Valla should
have influenced Nicholas with whom he was on friendly terms in later years.

[2] Valla, 'Annotationes in N.T.', *Opera* p. 852; cf. G. Mancini, *Vita di L. Valla* (1891) p. 312.

[3] Valla, 'In errores Antonii Raudensis adnotationes', *Opera* p. 428 '(Seneca) . . . de ementitis
ad Paulum, et Pauli ad eum epistulis, in alio opere disputavimus'; cf. A. Momigliano,
'Contributo alla storia degli studi classici' [1] [1955] = *Storia e letteratura* 47, pp. 28 f. and
'Secondo contributo' [1960] = *Storia e letteratura* 77, pp. 106 f. with references to Giovanni
Colonna and Boccaccio.

[4] I follow with due reserve the chronology of J. Vahlen, *Laurentii Vallae Opuscula tria* (1869)
p. 58 (reprinted in *Opera omnia* II 184).

wrote bears the stamp of genius, there have been greater philosophers; but there was no greater scholar in the middle of the Quattrocento. Valla's most valuable legacy to future scholars was the rationalism and the mistrust of so-called authorities evident in all his work, but brilliantly summed up in a short rhetorical question, uttered towards the end of his life in 1455: 'An melior ullus auctor est quam ratio?'[1]

[1] 'Confutatio prior in Benedictum Morandum Bononiensem', *Op.* p. 448, l. 16; on Benedetto Morandi who rather stupidly attacked one of Valla's historical interpretations of Livy see Mancini, *Vita di L. Valla* p. 318.

IV

POLITIAN

THE radiant figure of Politian stands at the end of the Quattrocento. Angelo Ambrogini,[1] called Poliziano after Monte Pulciano, where he was born in 1454, came to Florence at the age of ten as an orphan and infant prodigy, and died there in 1494 shortly after his great patron and friend Lorenzo de' Medici who had been in power from 1469 to 1492 and whose death was passionately lamented by Politian in a Latin ode[2] and in a long moving letter to Jacopo Antiquario.[3] Politian's teachers were the remarkable Latin and Greek scholars of the Medicean circle, and he became tutor to Lorenzo's children, lecturing on Latin and Greek literature from 1480 onwards to a wide audience of Italian and foreign students.

Politian, whose personal charm seems to have been irresistible, represents the fifth and last generation of the productive Italian scholars of the Renaissance. We have compared[4] this living chain of freely associated scholars through five generations from Petrarch to Politian with the five generations of the great Alexandrians from Philitas and Zenodotus to Aristarchus. As ποιητὴς ἅμα καὶ κριτικός

---

[1] *Prosatori* pp. 867 f. Poliziano. Short biographical and bibliographical introductions by E. Garin. There is neither a new critical edition of his works nor a modern biography. We shall have to turn to the first collection of his *Opera* in the Aldine edition of 1498; later editions (Flor. 1499, Lugd. 1528, Lugd. 1537–9, reprinted 1971, Bas. 1553) are reprints with some additions. The most important collection is still *Prose volgari inedite e poesie Latine e Greche edite e inedite* di A. Poliziano, raccolte e illustrate da Isidoro Del Lungo (1867). All these editions and collections are put together by Ida Maier: A. Politianus, *Opera omnia* (I: Scripta in ed. Basilensi anno MDLIII collecta; II: Opera ab Isidoro Del Lungo edita. Florentiae anno MDCCCLXVII; III: Opera miscellanea et epistulae), Rist. anast. 1970/1. E. Garin, op. cit. pp. 869–925 reprints only a very small collection of two opuscula and three letters. A. Poliziano, *Miscellaneorum centuria secunda*, Edizione critica per cura di Vittore Branca e Manlio Pastore Stocchi, 4 vols. (Firenze 1972). *Mostra del Poliziano nella biblioteca Medicea Laurenziana* (1954), Manoscritti, libri rari, autografi e documenti, 12 plates, Catalogo a cura di A. Perosa, is of fundamental importance, as it contains an almost complete 'bibliographie raisonnée' by the greatest expert; we expect from the same author a commentary on the *Miscellanea*. See also below, p. 44 n. 7 with reference to the *Atti del IV convegno* which contains some other important articles on Poliziano. On the letters, published in various places and unpublished ones, see A. Campana, 'Per il carteggio del Poliziano', *La Rinascità* 6 (1943) 437–72; we must still look forward to this *carteggio* which will be also an inexhaustible source for his scholarship and polemics against his critics.

[2] Del Lungo, p. 274 *Odae* XI.   [3] *Prosatori* pp. 886 ff.   [4] *History* [I] 233; cf. 90, 170.

Politian was nearer to Philitas than anyone else at any time. He was first of all a true Italian poet in the vernacular;[1] but in poems like *Orfeo* or *Giostra* the spirit of the Italian people and the formal power of antiquity are intimately connected. The *Orfeo* is supposed[2] to have been one of the first Italian plays in which the tradition of 'sacra rappresentazione' was transferred to a secular drama; a few songs and choruses set to music—now lost—opened the way to the modern opera.

This great poet was also a passionate seeker after the factual information that he found necessary for understanding and interpreting ancient poetry.[3] His *ardor eruditionis* brought him a comprehensive knowledge not only of the Latin, but also of the Greek world. By far the greater part of his work was, of course, devoted to Latin: to the poets not only of the Augustan age, but also and especially to those of the 'Silver Age', as they were termed from Erasmian times on, Statius,[4] Lucan, Seneca, Ausonius; and his labours extended even to the Roman legal texts, of which he had the most important manuscript at his disposal.[5] None of the great Italian scholars before Politian had been a Ciceronian; even Valla used to accept 'quod magnis auctoribus placuit'. The movement called Ciceronianism[6] was founded and propagated by distinguished schoolmasters, Gasparino da Barzizza (d. 1431) and Guarino da Verona (d. 1460). When Politian was reproached by one of their followers for his un-Ciceronian style he rebuked him in his programmatic letter to Paolo Cortese about the 'apes of Cicero': '"Non exprimis", inquit aliquis, "Ciceronem". Quid tum? Non enim sum Cicero. Me tamen, ut opinor, exprimo.'[7] As a poet, he was also a very sensitive stylist in prose, and was not afraid to mix classical Latin, if necessary, with rare and archaic words.

Politian began to learn Greek at the age of ten. When he was sixteen he was able to write Greek verse, and at eighteen he translated Books II–V of the *Iliad* into magnificent Latin hexameters, leaving Bruni's and Valla's attempts far behind. Politian was the first Western scholar who could rival the Greek immigrants in knowledge of the

---

[1] *Stanze* ed. V. Pernicone (1954); see *Geschichte der Textüberlieferung* II (1964) 529 ff.

[2] *Oxford History of Music* IV (1968) 786.

[3] See *Misc.* ch. IV 'qui poetarum interpretationem suscipit', the essential chapter on interpretation.

[4] Stat. *Silv.* ed. A. Marastoni (1961) pp. lvi–xc A. Politian Silvarum emendator.

[5] *Misc.* ch. XLI and *passim*, see A. Perosa, *Mostra del Poliziano*, Catalogo n. 47 on the so-called codex Pisanus of the Pandects; cf. nn. 48–50 and 82.

[6] R. Sabbadini, *Storia del Ciceronianismo* (1885).

[7] *Epp.* VIII 16 (Politiani epistolae, Amstelodami 1642, pp. 307 f.); cf. Th. Zielinski, *Cicero im Wandel der Jahrhunderte*[2] (1908) p. 425.

ancient Greek language; this distinguishes him from all his Italian predecessors. He may also have been the first Italian to try to emend a Greek text and to supply words missing in the manuscripts.[1]

Politian lectured on Homer, Hesiod, and Theocritus. The introductions to his lectures were not collections of dry biographical and bibliographical information, but genuine poems in Latin hexameters, which he called 'Silvae' in honour of Statius;[2] the *praelectio* to Homer (1485),[3] for instance, was based on an exact knowledge of the ancient βίοι ʽΟμήρου,[4] but all this learning was transformed into poetical beauty;[5] It was the first time Greek, as opposed to Latin, epics had been eulogized by an experienced Western scholar poet. As in Latin literature, so also in Greek he turned to post-classical poets, to Theocritus, Callimachus, and the epigrammatists of the *Anthology*; he even ventured to compose a fair number of epigrams in Greek.[6]

Politian did not complete any edition in his short life. Our sources are his marginal notes in his books, the excerpts in his 'Zibaldoni autografi', and his printed *Miscellanea*. He was a voracious reader, not only of the books he owned, but also of the many available to him from the Florentine libraries, and he was indefatigable in jotting down his notes.[7] Of the books he used scarcely one is without any traces of his study. He was apparently the first to make complete collations of manuscripts[8] and carefully noted the readings in his own copy; he seems even to have started to use special sigla[9] for manuscripts, which slowly became familiar to later generations. Politian insisted more than anyone else on the importance of a knowledge of the best manuscripts as a defence against the rash conjectures of his contemporaries.[10] There must be innumerable notes of his still to be gathered from the margins

[1] Cf. below, p. 45 f.

[2] 'Silva in scabiem' was apparently of less importance than the four introductory 'Silvae'; a copy was only recently discovered and published by A. Perosa in *Note e discussioni* 4 (1954) with an important introduction on the tradition of Politian's Latin poems.

[3] *Prose volgari . . . e poesie Latine e Greche . . .*, raccolte e illustrate da Isidoro Del Lungo (1867) pp. 333–68; p. 360 praise of Homer.

[4] *History* [1] 11.

[5] In his 'Oratio in expositione Homeri', *Opp.* (Basel 1553) pp. 474–92 he adapted [Plut.] *De Hom. poesi*.

[6] Poliziano, *Epigrammi Greci*, Introduzione, Testo e Traduzione di A. Ardizzoni, Biblioteca di Studi superiori 12 (1951); cf. E. Bignone, *Studi di filol. class.* N.s. 4 (1925) 391 ff.

[7] See A. Campana, 'Contributi alla biblioteca del Poliziano', *Il Poliziano e il suo tempo, Atti del IV convegno internazionale di studi sul rinascimento*, Firenze 1954 (1957) pp. 174 ff.

[8] G. Billanovich, *Journal of the Warburg and Courtauld Institutes* 14 (1951) 178. On the occasional deficiencies of his collations see Pasquali, *Storia* pp. 74 f.

[9] Campana (above n. 7) p. 202.

[10] S. Timpanaro, *La genesi del metoto del Lachmann* (1963) pp. 4 ff.

of his manuscripts and early printed books;[1] but this would be a difficult task as his books[2], scattered after his early death, are now preserved in various libraries,[3] and no inventory of them exists.

In his 'Zibaldoni autografi'[4] he seems to have made a start at collecting excerpts and textual notes, of which an exceptionally large number refer to the Pandects, as a preparation for the *Miscellanea* which were planned on a grand scale. The first centuria of his *Miscellanea*, splendidly printed in 1489 and dedicated[5] to Lorenzo de' Medici, contains his final collection of marginal notes to which he added discussions of grammatical, chronological, and antiquarian topics,[6] reconstructions of lost Greek originals from quotations and Latin imitations—a particularly bold step—and translations from Greek into Latin with textual criticism. Politian[7] was the first to put together a few fragments and testimonia of Callimachus' *Hecale*[8] and fragments of the Callimachean poem Βερενίκης πλόκαμος translated by Catullus (66).[9] It is possible, too, that he tried his hand at those parts of the *Aetia* so often quoted in later Greek and Latin literature.[10] In his circle the text of Callimachus' *Hymns* was most eagerly copied,[11] and he himself translated the *Bath of Pallas*.[12] In his notes to this hymn he confessed 'that he was not afraid to correct small corruptions of the original';[13] but in one passage (line 136), where he was thought to have divined the true Callimachean text,[14] we now know that he was completely astray. Only the concluding word was preserved in Politian's Greek manuscript, and in his Latin

---

[1] Perosa, *Mostra del Poliziano*, paid special attention to the annotated copies; a notable addition is given by M. Gigante, 'De A. Politiani notis in Cic. *de or.*', *Charisteria F. Novotny oblata* (1962) pp. 62 ff.

[2] Campana (above p. 44 n. 7) p. 174; cf. p. 178 'molto resta ancora da scoprire.'

[3] I had the opportunity of inspecting copies with Politian's notes in the Bodleian Library in Oxford and the State Library in Munich. Cf. also below, p. 136 with n. 4.

[4] Carmine di Pierro, 'Zibaldoni autografi di A. Poliziano', *Giornale storico della letteratura Italiana* 55 (1910) 1–32.

[5] In the dedication he spoke also of the title; was he the first to use the word to denote a literary work of mixed contents?

[6] A striking example of Politian's use of a rare Greek text (Johannes Lydus) printed about three centuries later is given by J. Bernays, *Gesammelte Abhandlungen* II (1885) 331 ff.

[7] Call. II p. xliii.   [8] *Misc.* ch. XXIV.

[9] *Misc.* chs. LXVIII and LXIX.

[10] V. Branca, 'La incompiuta seconda centuria dei "Miscellanea" di A. Poliziano', *Lettere Italiane* 13 (1961) 137–77 with 7 plates; pp. 149 and 161 on the *Aetia*. See the edition of the *centuria secunda* (1972) above, p. 42 n. 1.

[11] Call. II p. lxvii.

[12] *Misc.* ch. LXXX; reprinted by Del Lungo 1867 (see above, p. 42 n. 1) pp. 529 ff.

[13] Cf. above, p. 44; in the apparatus criticus of my edition of the *Hymns* one can easily see Politian's corrections.

[14] Del Lungo, op. cit. p. 538; cf. L. Ruberto, 'Studi sul Poliziano filologo', *Riv. fil. cl.* 12 (1883) 224.6.

translation he supplied the pentameter according to the sense he expected. This seemed to be confirmed by the text F. Robortello used in his edition of Callimachus' *Hymns* (1555), where the Greek line agreed completely with Politian's Latin one. But unfortunately, it is the other way round. Robortello's manuscript is one of the interpolated manuscripts of the sixteenth century in which all the gaps of the archetype are filled by modern supplements, and the one in question is nothing but a poor translation of Politian's Latin into Greek. That both Politian's guess at the contents of the line and all the Greek words are wrong is now proved by another group of manuscripts, unknown to him, in which four syllables of the beginning of the pentameter are preserved. Yet even when this modern error is discounted Politian's achievements as a scholar poet were considerable enough to earn him lasting fame. Death—a central theme in his own poetry—came to him at the age of forty in the horrible year of the French invasion, Piero di Lorenzo's flight, and the end of the Medici regime. His advance into the field of Greek epic poetry was at the time no more than an episode, because no Italian scholar of the next generations followed him; yet in due course, as we shall see, the French came to occupy the ground he had reconnoitred.

V

GENERAL ACHIEVEMENTS OF
SCHOLARSHIP IN ITALY AND ITS SPREAD
INTO TRANSALPINE COUNTRIES

OUR survey has so far been confined to the leading Italian humanists
of the fourteenth and fifteenth centuries, chiefly in Florence and Rome.
This has meant, of course, simplifying the historical process and leaving
considerable gaps; for some achievements such as the institution of
public libraries, the collection of antiquities, and the foundation of
schools and academies, cannot be attributed directly to these leading
humanists, though they were made possible and inspired by their
labours. In this chapter we try to fill these gaps and to describe the
results of Italian scholarship in general[1] and its effect on other European
countries.

We have stressed the unique importance of the book for scholarship;[2]
but there was no 'tyranny' of the book at any time in Greek and
Roman antiquity, no cult of the written word merely because it was
written, as there had been in the Oriental and medieval worlds. The
bent of the Greek mind had been critical, and a similar critical attitude
is found again in the Italian Renaissance.

The scholar poets were in need of good texts; we have seen how they
tried to find and to correct them. But not only this: there was a conscious
tendency from the beginning to bring the book from the monastic and
cathedral libraries, first into private hands, and finally into great new
libraries more or less accessible to everyone interested in literature and
learning.

We have followed the travels of many scholars through Central and
Western Europe as they hunted for Latin manuscripts[3] from Petrarch's

---

[1] Cf. R. Sabbadini, 'Il metodo degli umanisti', *Bibliotechina del Saggiatore* 3 (1920); for textual
criticism see the chapters on Valla and Politian.

[2] *History* [I] 17, 102 f.

[3] See above, pp. 9 ff. and *passim*, esp. pp. 30 ff. See also S. Prete, 'Die Leistungen der
Humanisten auf dem Gebiete der lateinischen Philologie', *Philologus* 109 (1963) 258 ff. with
new material from manuscripts.

days on. But there was also much travelling between West and East in quest of Greek manuscripts, though none of the leading scholars took part in it. The fame of Giovanni Aurispa[1] and Francesco Filelfo does not rest on their scholarship, but on their highly successful trading in Greek codices. The harvest of Aurispa's two trips to the Orient in 1405–13 and 1421–3 was unique. He is said to have rescued altogether about 300 Greek manuscripts from the East, among which the outstanding pieces were: a codex of the *Iliad*, containing text and scholia with substantial passages of Aristarchean material,[2] which went to the library of St. Mark in Venice (Codex Venet. Marc. 454); the tenth- or eleventh-century codex with seven plays of Sophocles, six of Aeschylus, and Apollonius Rhodius' *Argonautica*, now in the Laurentian library in Florence (Codex Laurentianus XXXII 9);[3] possibly the archetype[4] subsequently lost of the great collection of Greek hymns (Homer, Callimachus, Orpheus, Proclus) and the codex of the *Greek Anthology* of epigrams (Codex Palatinus 23).[5] Filelfo, an attaché of the Venetian embassy in Constantinople from 1420 to 1427, returned from his post with manuscripts of about forty Greek authors.[6] We have mentioned[7] how agents of the Medici were instructed by Niccoli about the hiding-places of Latin classics; Lorenzo's principal agent in the East was Janus Lascaris (1445–1535),[8] who brought in 1491 about 200 Greek manuscripts from Mount Athos to Florence; even the diary he kept during his travels is preserved.[9] After his return he produced, between 1494 and 1496, five magnificent first editions in Greek capitals. After the end of the Medici regime in Florence he went to Paris[10] and later also to Rome under the second Medicean pope Leo X.

In the seeking out and collecting of classics there was a mixture of

---

[1] Giovanni Aurispa, *Carteggio* ed. R. Sabbadini, Fonti per la storia d'Italia 70 (1931); cf. Sabbadini, 'G. Aurispa, scopritore di testi antichi', *Historia* 1 (1927) 77–84. See also Aurispa's Greek version of Cardinal Cesarini's Latin speech with which he addressed the Eastern legates at the Council of Basle in 1434, published with important notes by B. Wyss, *Mus. Helv.* 22 (1965) 1 ff.

[2] See *History* [1] 213 f.

[3] The representative Catalogue *Mostra della Biblioteca di Lorenzo nella Biblioteca Medicea Laurenziana* (1949) p. 18 unfortunately relapsed into the error that all our extant manuscripts of Aeschylus are derived from the Mediceus.

[4] Call. II, pp. lxxxi f.   [5] Call. II, p. xciii.

[6] A. Calderini, *Studi ital. fil. class.* 20 (1913) 204 ff.

[7] See above, p. 30.

[8] See Call. II, pp. lxvi f.; l.8 from the bottom read 'quattuor' instead of 'quinque'. I am sceptical about the attribution of Theognis MSS. and the first edition to Ianus Lascaris by Douglas C. C. Young, 'A Codicological Inventory of Theognis MSS.', *Scriptorium* 7 (1953) 3 ff.

[9] H. Hunger, *Jahrbuch der österreich. byzantinischen Gesellschaft* 11/12 (1962/3) 117.

[10] B. Knös, *Un Ambassadeur d'hellénisme, Ianus Lascaris, et la tradition greco-byzantine dans l'humanisme français* (Uppsala—Paris 1945); cf. below, p. 61.

free enterprise of individual scholars and conscious organization by mighty patrons; it was the same with the storing of these books in libraries.[1] Petrarch, Salutati, and Niccoli built up remarkable private libraries;[2] others like Boccaccio, Poggio,[3] and Politian[4] were relatively poor scholars who tried hard to do the same, but without great success. Only the power and wealth of princes, popes, and cardinals, stirred by the enthusiasm of the great scholars, were able to found or to enlarge great treasure-houses for the preservation of Latin and Greek manuscripts. Petrarch, always in advance of his age, had had the idea of leaving his books to the Republic of Venice as the nucleus of a future public library. But Florence took the lead under the Medici: Cosimo (1389–1464) and Lorenzo (1469–92) not only built up their palace library, but also contributed generously to the public library of the Dominican convent of S. Marco[5] and the Benedictine Abbey of Badia in Fiesole. In the end most of their books came to rest after many adventures in the splendid building in the court of S. Lorenzo, which Giulio de' Medici as pope Clement VII had commissioned from Michelangelo in 1523, though it was not inaugurated and thrown open to the public until 1571. Tommaso Parentucelli, growing up under the strong influence of the Florentine circle, had passionately tried as a poor monk to buy books and have them copied; on the papal throne as Nicolas V (1447–55), he initiated the famous department of classical (especially Greek) manuscripts in the Vatican library. As the protector of the humanists he gained the support of the powerful cultural movement for the Holy See. Cardinal Bessarion[6] in 1468 presented about 800 manuscripts—amongst them nearly 500 Greek ones—to the Republic of Venice which had already profited so much from Aurispa's and Filelfo's travels. Other places, such as Milan, Pavia,[7] and Naples,

---

[1] *Handbuch der Bibliothekswissenschaft*, 2. Aufl. III 1 (1955) 499 ff. A. Bömer—H. Widmann, 'Renaissance und Humanismus'.

[2] See above pp. 12, 27, 30; cf. G. Billanovich, 'Les Bibliothèques des humanistes italiens au XIVᵉ siècle', *L'Humanisme médiéval dans les littératures Romanes du XIIᵉ au XIVᵉ siècle*, *Actes et colloques* 3 (1964) 196–214.

[3] Walser, *Poggius* pp. 104 ff.

[4] Campana, 'Contributi alla biblioteca del Poliziano' (above, p. 44 n. 7) p. 174 'studioso e professore, cioè povero'.

[5] See above, pp. 23, 30.

[6] See above, p. 37 (with n. 8) on the mistaken assumption that Valla's biblical criticism was influenced by Bessarion.—An inventory of his library was published by H. Omont, *Revue des bibliothèques* 4 (1894) 129–87; two hitherto unedited catalogues will be published by L. Labowsky, see Annual Report 1965/6 *Proceedings of the British Academy* 52 (1966) 25. *Cento Codici Bessarionei*. Catalogo di Mostra a cura di Tullia Gasparrini Leporace ed Elpidio Mioni (1968) with the text of the presentation.

[7] E. Pellegrin, *La Bibliothèque des Visconti et des Sforza* (Paris 1955)

and smaller ones, such as Urbino or Ferrara, were ambitious enough to build up their own libraries. Political and social conditions in Italy favoured this pleasant variety, in contrast to the centralization of Hellenistic Alexandria, which had been without rivals until the rise of Pergamum.[1]

The contents of Italian libraries were augmented, after printing had been introduced into Italy by two German printers in 1465 and Italian printers and Greek immigrants began to produce their *editiones principes*[2] of Latin and Greek classics. Their chief aim was to print books that were beautiful in type and size and paper (which had almost replaced vellum by that time) and to diffuse the knowledge of the beauty and the wisdom of the classics among a wider public. But printing at first made no real contribution to the improvement of the texts, as it was the practice to send individual manuscripts, and often not the best ones, to press unedited; in fact, to counterbalance the advantage of printing, there was now the danger that bad texts might be reprinted and stored in large numbers in private and public libraries.

It would have been absurd if an age of the greatest artists had taken an interest only in classical books, and not in the remains of ancient monuments. We need not be surprised that the generation of Niccoli and Poggio[3] began to collect epigraphical and archaeological material in Italy. Poggio's slightly younger contemporary Flavio Biondo[4] (1392–1463) was already able to include in his *Roma triumphans* (1456–60) many Roman public, private, religious, and military antiquities; this classification remained fundamental for centuries. His *Roma instaurata* (1440–63), after describing the city, deals with the restoration of the ancient monuments; his *Italia instaurata* (1456–60) gives a topographical survey of antiquities in the whole of Italy. The stimulating effect of these books was such that they were soon superseded by further antiquarian and topographical research. When Biondo turned to write history, there was no need, of course, to write ancient history—

---

[1] *History* [1] 98 ff. and 235 ff.

[2] Sandys II 102 ff. with chronological lists; of special importance is the collection of prefaces to the first editions by B. Botfield, *Praefationes et Epistolae* (1861) R. Hirsch, *Printing, Selling and Reading 1450–1550* (1967); see especially Chapter I: 'From script to printing' pp. 1–12; pp. 138 ff. on early printed editions of classics and humanists. On the Venetian printers Aldus Manutius and his family see below, p. 56; see also *La stampa greca a Venezia nei secoli XV e XVI*. Catalogo di Mostra a cura di Marcello Finazzi (1968).

[3] See above, p. 33.

[4] A. Momigliano, 'Ancient History and the Antiquarian', in *Contributo alla storia degli studi classici*, Storia e Letteratura 47 (1955) 75 ff.; B. Nogara's introduction to 'Scritti inediti e rari di Biondo Flavio', *Studi e Testi* 48 (1927) gives a new and most valuable biographical sketch (clxxxiii pp.) together with a number of new texts.

the ancient historians had done that. He concentrated on the history of post-ancient times from 410 to 1441: *Historiarum ab inclinatione Romani imperii Decades* in 42 books.[1]

Cristoforo de' Buondelmonti[2] and the remarkable Ciriaco di Ancona[3] (1391-1455) also visited the countries of the Byzantine Empire; Ciriaco recorded remains of classical antiquities, copied inscriptions, and made drawings of the monuments. His lively enthusiasm for his subject and ignorance of his own shortcomings gave him the courage to undertake all this; and posterity has every reason to be grateful for the substantial contribution he made to the knowledge of ancient relics, some of which were afterwards lost. Ciriaco's work was continued by a gifted and faithful, though eccentric, disciple,[4] Felice Feliciano of Verona (1433-79?), an epigraphist and calligrapher, whose merits have been duly acknowledged by Mommsen.[5]

Turning from the eastern travellers back to Rome, we find the strange figure of an antiquarian who took the rediscovered Roman world so seriously that he tried to revive its customs and to follow the prescriptions of its writers on daily life, Pomponio Leto (1428-97).[6] His house on the Quirinal was filled with antiquities, coins, medals, inscriptions; he did not care for Greek, but strove for perfection in Latin and for a graceful Latin script. About 1460 he founded his so-called Roman Academy, and spoke almost daily to a devoted audience, which included students from many foreign countries and made him famous all over Europe. Unfortunately, celebrations of pagan rites and secret meetings in the catacombs roused the suspicion of the Church; harmless though his eccentricity seems, Leto together with the chief members of the Academy, was imprisoned and even tortured by order

---

[1] Denys Hay, 'Flavio Biondo and the Middle Ages', *Proc. Brit. Acad.* 45 (1959) 97 ff.

[2] Cf. above, p. 29 and esp. n. 1 (E. Jacobs).

[3] R. Sabbadini, *Enciclopedia italiana* 10 (1931) 438 s.v.; B. Ashmole, 'Cyriac of Ancona', *Proc. Brit. Acad.* 45 (1959) 25-41 confines himself mainly to the problem of Cyriac's fidelity when recording antiquities, but gives many helpful references, including some to future publications, p. 26.1. The most extensive autograph, revealing his interests and activities, is his diary of 1436, published by P. Maas, 'Ein Notizbuch des Cyriacus von Ancona', *Beiträge zur Forschung*, Studien und Mitteilungen aus dem Antiquariat Jacques Rosenthal, Folge I, Heft 1 (1923) 5-15; see also Wardrop, *The Script of Humanism* (above p. 13 n. 6) pp. 14 ff. on Ciriaco and Felice Feliciano.

[4] C. Mitchell, 'Felice Feliciano Antiquarius', *Proc. Brit. Acad.* 47 (1961) 197 ff.

[5] *CIL* v (1872); new evidence enabled Mitchell, op. cit. pp. 211 ff. to correct some of Mommsen's mistakes and to improve on our knowledge of Feliciano's achievements.

[6] Vladimir Zabughin, *Giulio Pomponio Leto*, 2 vols. (Grottaferrata and Rome 1909-12, translated from the Russian original edition); cf. the excellent section on P.L. in J. Wardrop (above p. 13 n. 6) pp. 20-3. See also J. Delz, 'Ein unbekannter Brief von Pomponius Laetus', *IMU* 9 (1966) 417 ff., for new material on Leto's process and his teaching activity in Venice.

of Pope Paul II, and the Academy was dissolved in 1467. But Paul's successors tried honourably to make amends for this anticipation of the shameful persecutions of the counter-reformation, and the Academy, restored by Sixtus IV, flourished again under Leo X.

Leto's Academy may be regarded as an example of the development of a scholar's privately arranged lectures into an organized school which lived on after its founder's death; we may compare the development of personal into public libraries. Although the humanistic movement had by its nature a strong educational tendency,[1] it is not possible here to give a full history of school-teaching; we have only occasionally been able to mention the teaching activity of scholars and what follows must be confined to the relation of school-teaching to scholarship. It is vital to keep up this connection from both sides: the schools need the guidance of scholarship, and scholarship depends on continuity of language teaching and the explanation of texts in schools. It was, of course, a slow process to reform and reorganize the monastic, cathedral, city, and private schools and the old universities in the humanistic spirit; some died out, and new foundations were established in various places.

We noticed that in the later Hellenistic age (about 100 B.C.) the list of teachers was headed by the γραμματικοί[2] and that grammar remained the first of the literary arts for all times. The late medieval teaching grammar of Alexander of Villedieu (about A.D. 1200) and similar school-books were still widely used in the fourteenth and fifteenth centuries; based in general on Priscian they taught late classical ' medieval Latin. When Erasmus started the fight against the barbar:.ns (about 1484?), he could not refrain from ironical mockery of Alexander and grammatical textbooks like his.[3] The first to write grammatical rules on the basis of the new study of classical literature seems to have been Guarino da Verona, the great educationalist, in his *Regulae grammaticae* (1418); but only fifty years later the *Rudimenta grammatica* of Niccolò Perotti (1468) succeeded in ousting the antiquated Latin grammars.

Greek grammars were in due course produced by the Greek immigrants who taught ancient Greek in Italy. Salutati, as we have related, succeeded in bringing Chrysoloras from Constantinople to Florence.[4]

---

[1] Cf. G. Saitta, *L'educazione dell' umanesimo in Italia* (1928).

[2] *History* [1] 253.

[3] 'Die Wandlungen der *Antibarbari*', *Gedenkschrift zum 400. Todestag des Erasmus von Rotterdam* (1936) 64 = *Ausgewählte Schriften* 203.52.

[4] See above, p. 27.

For his pupils—some of them very distinguished scholars—Chrysoloras wrote his Ἐρωτήματα τῆς Ἑλληνικῆς γλώσσης, for which he naturally used the Τέχνη γραμματική of Dionysius Thrax,[1] thus enabling that tiny volume, already fundamental for the Eastern world, to conquer the West. Guarino, who had followed Chrysoloras for five years to Constantinople and returned with more than fifty Greek manuscripts, translated excerpts from the Ἐρωτήματα into Latin. His son Battista Guarino expressed his father's main ideas[2] in a treatise *De ordine docendi et studendi*, written in 1459.[3] After a long passage on Latin he stresses the importance of the study of Greek: 'Frequentanda erit in primis Graecarum literarum lectio.' It was by no means usual in the middle of the Italian Quattrocento to insist on the knowledge *utriusque linguae* for educational purposes; as to scholarship, no Italian scholar managed to be creative in both fields, except Politian[4] after 1480 for a few years. The first printed Greek grammar, and actually the first book ever printed in Greek, was the Ἐρωτήματα of Constantinus Lascaris (1476, reprinted 1966); the Aldine edition of 1508 included an Appendix dealing especially with the problem of the pronunciation of ancient Greek, which Janus Lascaris and other Greeks had started to discuss long before it became a popular topic through the monographs of Reuchlin and Erasmus.[5]

Latin kept its predominant position.[6] Though the elementary rules were taken for granted, the question of Latin 'style', that is of vocabulary and phraseology, was hotly debated.[7] It is understandable that school-masters tended to solve the whole problem by encouraging the exclusive imitation of one great writer, Cicero, who had enjoyed a unique reputation since Petrarch's day; Ciceronianism was thus founded and propagated by these circles,[8] not by the great scholars. It was taken up by scholars of a higher order in the first third of the sixteenth century, when cardinal Bembo, papal secretary under Leo X, was its most

---

[1] See *History* [1] 266 ff.

[2] *De modo et ordine docendi ac discendi* (see note 3) p. xi b; cf. viii b 'neminem posse absque literarum Graecarum scientia in hac versuum doctrina fundamenta ... introspicere.'

[3] I use a printed edition (Strassburg 1514) with the title *De modo et ordine docendi ac discendi* on the title-page. W. H. Woodward, *Vittorino da Feltre and other Humanist Educators* (1897) pp. 159–78 (his English translation is based on another printed text).

[4] See above, pp. 42 and 46.

[5] I. Bywater, *The Erasmian Pronunciation of Greek and its Precursors* (London 1908).

[6] See S. Prete, 'Die Leistungen der Humanisten auf dem Gebiete der lateinischen Philologie', *Philol.* 109 (1965) 259–69, with very important examples and references, esp. on Guarino da Verona.

[7] See above, pp. 35 f.

[8] Cf. above, p. 43; see also Sandys, *Harvard Lectures* (1905) pp. 145 ff., esp. 149 ff.

distinguished representative; and finally in 1528 Erasmus in his dialogue *Ciceronianus*[1] gave a witty caricature of extreme Ciceronianism and showed the way to a moderate and correct adaptation of Cicero's eloquence.

The ancient texts needed careful interpretation; it seems that the writing of commentaries grew slowly out of viva-voce explication in schools. We found no evidence that in early Hellenistic times pupils, say of Zenodotus, noted down his oral exegesis of Homer and handed it on to posterity; we simply had to assume a tradition of this kind;[2] for we know of no written commentaries before Aristarchus, that is, before the middle of the second century B.C.

In the Italian Renaissance, too, they came on the scene relatively late; the reason was not only that the working out of complete commentaries naturally followed after that of editions of texts, interpretations of selected passages, and monographs, but also that the ancient scholiasts, as far as they were extant or rediscovered, could be used,[3] sometimes with modern additions.[4] Nevertheless, it seems that as early as Petrarch's day, his friend and admirer, the Augustinian monk Dionigi de' Roberti,[5] composed for his lectures a sort of commentary on Valerius Maximus, a favourite of the Renaissance, and on a few Roman poets, the explanatory notes of which he then had copied and dedicated to cardinal Giovanni Colonna. Gasparino da Barzizza, a renowned teacher, undertook sketchy expositions of a few of Cicero's rhetorical and philosophical writings and of his letters. A notable advance was made when in the seventies Niccolò Perotti, cardinal Bessarion's protégé, the translator of Polybius and the author of a new popular Latin grammar,[6] collected material for a bulky encyclopedic

---

[1] There is a separate edition of the revised text by I. C. Schönberger (1919); it is a pity that he never published the promised commentary.

[2] *History* [I] 212 ff., cf. p. 108.   [3] See above, p. 3.

[4] G. N. Knauer, 'Die Aeneis und Homer', *Hypomnemata* 7 (1964) 77.2. The additions to Servius' commentary on Virgil, written in 1459 in Ferrara, might be due to Guarino's school. Battista Guarino in his book *De ordine docendi* p. xia (see above, p. 53) said 'explanationes quoque librorum scribere vehementer conducet, sed tamen magis si sperabunt eas in lucem aliquando credituras.'

[5] Cf. above, p. 11. The short survey 'Commentare zu Klassikern', Voigt II 387 ff., is still useful; but now, with the help of P. O. Kristeller, *Catalogus translationum et commentariorum* (Latin commentaries on ancient Greek and Latin authors up to the year 1600) I (1960), someone should investigate further in Italian libraries and try to write a history of the early commentaries. See F. Simone, 'Il Rinascimento Francese', *Biblioteca di Studi Francesi* I (1961) 16, 18, 19, 23 on Dionigi (and others). Cf. also *Der Kommentar in der Renaissance*, hrsg. v. A. Buck u. O. Herding., Kommission f. Humanismusforschung, Mitteilungen, I (1975).

[6] See above, p. 52. For a list of his writings see N. Perotti's version of the *Enchiridion* of Epictetus ed. R. P. Oliver (Urbana, Ill. 1954).

commentary on Martial, which was first published after his death in 1489 under the title *Cornucopiae* and often reprinted. There were no ancient scholia on any of these works. That Pomponio Leto 'produced . . . commentaries on the whole of Virgil, including the minor works (1487-90)', seems to be a traditional mistake;[1] Leto's 'notes on Virgil' are just one of the very rare survivals of the practice by which pupils took down the lectures of their master and published them afterwards.[2] With some relief we also learnt that Giovanni Tortelli,[3] the first Vatican librarian, never wrote a separate running commentary on Juvenal, as used to be supposed, but that in choosing passages for explanation in his work on *Orthography*[4] he simply took more from Juvenal, his favourite poet, than from other writers. Not until the very end of the century do we find a scholar, Filippo Beroaldo the elder (1453–1505), deliberately concentrating on the interpretation of Latin authors;[5] he became one of the few distinguished teachers of classics in the university of his native city Bologna, the famous seat of the legal learning. He may have been influenced to some extent by his great friend and correspondent Politian, who more than anyone insisted on the necessity of interpretation and whose *Miscellanea* provided even better examples of it than Valla's *Adnotationes*. Codrus Urceus, who taught Greek in the same university, said of Beroaldo 'per excellentiam quandam . . . commentatorem Bononiensem . . . legem . . . commentandi non servasse modo . . . sed prope constituisse'. If he was indeed the first to establish the 'lex commentandi', he had followers from the beginning of the sixteenth century onwards who were by no means afraid to comment also on the greatest Roman classics[6] and so to continue the work of the ancient scholiasts.

Despite the development we have just described at Bologna, the old

[1] So for instance Sandys II 93; on Leto see above, pp. 51 f.

[2] See Zabughin, *Pomponio Leto* I 264 (above p. 51 n. 6) and esp. Zabughin, *Virgilio nel Rinascimento Italiano* I (1921) 190 f., on the manuscripts he found in the Bodleian and Vatican libraries.

[3] Biographical and bibliographical details registered by R. P. Oliver, 'Giovanni Tortelli', *Studies presented to D. M. Robinson* II (1953) 1257–71; we are told some amusing anecdotes, but I was not able to discover any evidence for the story that Poggio threw a codex of Livy upon the head of his opponent Tortelli.

[4] *Commentaria grammatica de orthographia dictionum e Graecis tractarum* (written 1449, first printed 1471). On his quotations from Juvenal see E. M. Sanford, *TAPA* 82 (1951) 207 ff.— B. A. Müller, *Philolog. Wochenschr.* 50 (1930) 111 ff., offers some helpful remarks and references on punctuation in printed books of the Renaissance; cf. *History* [1] 178 f.

[5] 1476 Plin. *n.h.*, 1500 Apuleius, etc. On Beroaldo's method of commenting see K. Krautter, 'Philologische Methode und humanistische Existenz, Filippo Beroaldo und sein Kommentar zum Goldenen Esel des Apuleius', *Humanistische Bibliothek, Reihe* I: *Abhandlungen*, vol. 9 (1971) 37 ff.

[6] See Knauer (above p. 54 n. 4) pp. 64 ff. on Virgil.

Italian universities[1] did not play a decisive part in promoting the teaching side of Renaissance scholarship; as we shall see, more was achieved by new transalpine foundations. But there were important teaching institutions in Italy outside the universities. One of the earliest of these was the so-called 'Studio'[2] in Florence, founded in 1321, which dedicated itself to scholarship and literature, and invited nearly all the distinguished Italian and Greek scholars of the fourteenth and fifteenth centuries to give lectures. The 'Studio' was reformed in 1420; Politian was first its pupil, and afterwards one of its greatest luminaries. Another much younger institution was the 'Sodalitas $\Phi\iota\lambda\epsilon\lambda\lambda\acute{\eta}\nu\omega\nu$'[3] in Venice founded in 1500; scholars of foreign countries became members when they visited Italy (as did Erasmus, for instance, in 1508). Venice had its own Greek scriptoria, closely connected with those in Crete, and above all the great printing house of Aldus Manutius (1449–1515) and of his family. Aldus gratefully acknowledged himself a pupil of the younger Guarino,[4] who taught him Latin and Greek, and of another less famous Veronese scholar, Gasparino, an excellent Latin grammarian. So he was well equipped to become the editor as well as the printer of ancient and humanistic texts.[5] His Academy of Hellenists encouraged and helped him to produce in a single year, 1502, five first editions of Greek authors amongst a total of twenty-seven Greek *editiones principes* in twenty-one years.[6] He published the first collection of the writings of Politian of whom he was an enthusiastic admirer.

Beside the Venetian 'Sodalitas' and the Florentine 'Studio' new Academies arose in other cities. Pomponio Leto's highly individual Academy flourished and suffered in Rome, as we have seen. Poets and scholars were an ornament to the court of King Alfonso in Naples. Valla excelled in the learned debates in which the king took part,[7] but the Academy fostered by him seems to have been primarily an assembly

---

[1] A short survey of later medieval and early humanistic schools is given in F. Paulsen, *Geschichte des gelehrten Unterrichts* I³ (1919, reprinted 1960) 13–77. P. O. Kristeller, *Die italienischen Universitäten der Renaissance*, Schriften und Vorträge des Petrarca-Instituts Köln 1 [1953].

[2] Salutati, *Epist.* II 84 'legum et liberalium artium *Studium*'; see Isidoro Del Lungo, *Florentia* (1897) pp. 101 ff. on 'studio fiorentino', esp. Politian.

[3] D. J. Geanakoplos, *Greek Scholars in Venice* (1962) pp. 128 ff. on the foundation and the members of the 'Aldine Academy'.

[4] In the dedication of his edition of Theocritus (1495) to Battista Guarino (cf. above, p. 53).

[5] C. Dionisotti, 'Aldo Manuzio umanista', *Lettere Italiane* 12 (1960) 375 ff.

[6] On the Aldine Press see Geanakoplos (above n. 3) pp. 116 ff. and *passim*; on the five first editions of Greek authors in 1502 see A. A. Renouard, *Annales de l'imprimérie des Alde* (3rd ed. 1834, repr. 1953).

[7] See above, p. 36; cf. T. de Marinis, *La biblioteca Napoletana dei Rè d'Aragona*, 4 vols. (1947–52).

of poets, scholar poets of course such as Antonio Beccadelli, the author of the sensational *Hermaphroditus*, above all Giovanni Pontano (1426–1503)[1] whose elegant Latin verse and prose delighted Erasmus's ears, and finally Jacopo Sannazaro (1458–1530).

Florence became the meeting-place of East and West in. 1439 when a new Council was called together for the reunion of the churches. This led to the foundation under the protection of the Medici of the foremost of the Italian Academies. Georgios Gemistos, born in Constantinople, styling himself Plethon[2] (a synonym of Gemistos and in sound near to Platon), had elaborated his own philosophical system in the Neoplatonic tradition which had never died out in Byzantium. When he came to Florence as a member of the Council his Platonism made so deep an impression on Cosimo de' Medici (who had been in power since 1434) that he conceived the idea of setting up a freely organized Platonic Academy in his own city;[3] he even revived the custom of celebrating Plato's birthday by a banquet.[4] Luckily, he found in Marsilio Ficino (1433–99) a keen Platonist—he was said to keep a lamp in his room before the bust of Plato—who was able to translate and to interpret the Dialogues; the Latin Plato was completed in 1477 and printed in 1482, well before Musurus's Greek Aldine text of 1513. A Latin Plotinus, also by Ficino, followed in 1492. The availability of a complete text of Plato was probably more important than all the efforts to expound it anew. But efforts were made, and Ficino himself attempted a fresh interpretation in his *Theologia Platonica*,[5] supplemented by a book *De Christiana religione*; a fragment of his commentary on St. Paul's Epistle to the Romans is still extant. Ficino's nephew, Giovanni Pico della Mirandola (1463–94), the author of the programmatic *Oratio de hominis dignitate*,[6] lectured then in Florence on Plato and

---

[1] Ioannis Ioviani Pontani *De Sermone* libri sex, ed. S. Lupi et A. Risicato (1954) from an autograph in Vienna, with introduction and reference to Erasmus p. xiv; but Pontano also tried his hand on the text of Lucretius, Lucr. ed. Munro (1872³, pp. 6 and 11) who gives a first-hand survey of Lucretian scholarship in the Quattrocento. See also B. L. Ullman, 'Pontano's Handwriting', *IMU* 2 (1959) with 8 plates.

[2] F. Masai, *Pléthon et le platonism de Mistra* (Paris 1956); see esp. pp. 327 ff. 'Pléthon et les humanistes', 370 ff., and 384 ff.

[3] Nesca A. Robb, *Neoplatonism of the Italian Renaissance* (1935).

[4] So Ficino tells in the introduction to his translation of the *Symposium*; cf. R. Marcel, Marsile Ficin, Commentaire sur le *Banquet de Platon* (1956).

[5] Marsile Ficin, *Theologie platonicienne de l'immortalité des âmes*, 2 vols. Texte crit. établi et trad. par R. Marcel (1964–5).

[6] *L'opera e il pensiero di Giovanni Pico della Mirandola nella storia dell' Umanesimo*, Convegno internazionale, Mirandola 1963, Istituto nazionale di studi sul Rinascimento, Firenze 1965, contains various contributions on Pico's writings and on his influence on other countries. His *Carmina Latina* were discovered and edited by W. Speyer (1964).

on St. Paul. In no other Italian circle was the religious problem[1] of the relation between antiquity and Christianity so earnestly discussed. The Florentine way of reconciling them, however, was some kind of mystic symbolism.[2] This new Platonism had a universal influence on Spanish mysticism as well as upon the northern countries; St. Thomas More went so far as to translate into English a 'Life of Pico, Earl of Mirandola and a great lord of Italy'. It seems to be natural for fundamental Platonic ideas to undergo transformation in the course of history,[3] and to remain through all their metamorphoses a spiritual driving force. At the decisive moment Petrarch,[4] reversing the judgement of previous ages, had recognized Plato as 'philosophorum princeps', and as 'princeps' he was accepted by the Renaissance. But there was no strict uniformity of quality and interest amongst the nine members of the Florentine Academy. Politian was certainly one of the most notable members, and very fond of Pico, his junior by nine years, with whose praise he solemnly concluded his *Miscellanea*; but he never claimed to be a Platonist himself or indeed a philosopher at all, but only a 'grammaticus',[5] even when he was explaining a philosophic text or meditating about a system of scientific and scholarly doctrines in his *Panepistemon*.[6] Yet it is obvious from this book that he did not reject philosophy which from a systematic point of view included scholarship; he simply thought he himself belonged, not among the omniscient philosophers, but in the modest ranks of the scholars.

It was the Florentine Neoplatonism that had the strongest influence on the transalpine countries; but sporadic earlier contacts, official as well as private, had already prepared the ground for the reception of Italian scholarship in Central and Western Europe. Petrarch had come as an envoy from Milan in 1356 to Prague, where the Emperor Charles IV resided, and left a deep impression behind him. In 1361 the Emperor sent him copies of two documents on the privileges granted by Caesar and Nero to Austria; but after examining their language and style, he utterly denied their authenticity,[7] nearly a century before Valla's attack on the Constantine donation. Charles's chancellor,

---

[1] On the unproblematical attitude of Petrarch see above, pp. 11 f.

[2] On the different solution offered by Colet and Erasmus see below, p. 72.

[3] See *History* [1] 65.   [4] See above, p. 14.

[5] On γραμματικός see *History* [1] 157 f. A. Poliziano, *Le selve e la strega* per cura di Isidoro Del Lungo (1925) p. 222 'non ... philosophi nomen occupo ... nec aliud mihi nomen postulo quam grammatici'; cf. ['Lamia'] ibid. p. 226 'nomen vero aliis philosophi relinquero. Me ... grammaticum vocatote ...'.

[6] B. Weinberg, *A History of Literary Criticism in the Italian Renaissance* (1961) p. 3.

[7] *Lett. sen.* XVI 5 (vol. II 490 ff. Fracassetti), cf. E. H. Wilkins, *Life of Petrarch* (1961) p. 176.

Johannes von Neumarkt,[1] was an affectionate admirer of Petrarch, and in his intellectual circle we may find a modest analogy with some of the humanistic circles at courts in Italy. Rienzo, too, had found a refuge there in 1350, and the influence of his style on the language of the German chancellery is clearly perceptible.

But this early contact was not the real first-fruit of humanism in the north, only an episode without important consequences, which we might label as German prehumanism.[2] The second, more intensive contact was made nearly a century later by Enea Silvio Piccolomini (1405–64),[3] a pupil of Filelfo in Florence, again with the chancellery of the Emperor, now Frederick III, in Vienna (1442–55). Enea Silvio had taken part in the Council of Basle,[4] where for seven years after 1432 he watched the tragedy of its hopeless effort to reconcile Roman universalism with the particular interests of the nations. Convinced that only the leadership of the Church could save Europe, he became a priest and, in an extraordinarily short career, bishop of Trieste (in 1447), bishop of Siena (in 1450)—he was born in a village near Siena, Corsignano, now called Pienza after him—cardinal (in 1456), and finally in 1458 Pope Pius II: 'sum Pius Aeneas . . . fama super aethera notus' (Virg. *Aen.* 1 378 f.). All the time he was active as a gifted writer. He was not a great poet, though crowned as such by Frederick III, nor an eminent scholar. But he was widely learned and

---

[1] *Schriften Johanns von Neumarkt,* ed. J. Klapper, Vom Mittelalter zur Reformation 6.1, 2 (1930–2).

[2] E. Winter tried to demonstrate that an essentially religious movement of the fourteenth century in Bohemia was the link between Italian early humanism and 'Devotio moderna' which grew up in the Netherlands (see below, pp. 69 ff.): 'Die europäische Bedeutung des böhmischen Frühhumanismus', *Zeitschrift für deutsche Geistesgeschichte* 1 (1935) 233 ff. and especially 'Frühhumanismus. Seine Entwicklung in Böhmen und deren europäische Bedeutung für die Kirchenreformbestrebungen im 14. Jahrhundert', *Beiträge zur Geschichte des religiösen und wissenschaftlichen Denkens* 3 (Akademie-Verlag Berlin 1964) pp. 9, 169 f. Winter's views are strongly supported by H.-F. Rosenfeld, 'Zu den Anfängen der Devotio moderna', *Festgabe für U. Pretzel* (1963) pp. 239 ff. But this new hypothesis is not proved by the repetition of vague assurances; there seem to be no elements of scholarship in the Bohemian movement.

[3] *Opera omnia* (Basle 1551 [reprinted 1967] and 1571). 'Briefwechsel' ed. R. Wolkan, *Fontes rerum Austriacarum,* ii. Abteilung 61.62. 67.68. (1909–18) not yet complete; *Prosatori* ed. Garin (1952) pp. 661–87 (only a small excerpt from 'Commentarii rerum memorabilium'); *Enea Silvio Piccolomini. Papst Pius II. Ausgewählte Texte aus seinen Schriften* hrsg., übers. u. biographisch eingeleitet v. Berthe Widmer (1960); on Georg Voigt's monograph in three volumes see above, p. 17; G. Paparelli, *Enea Silvio Piccolomini,* Biblioteca di cultura moderna 481 (1950). E. Garin, 'Ritratto di E.S.P.', *La cultura filosofica del Rinascimento Italiano* (1961) pp. 38–59 with bibliography. Cf. also A. R. Baca, 'Enea Silvio Piccolomini's Verteidigung der Literatur', *Antike und Abendland* 17 (1971) 162 ff.

[4] *De gestis Concilii Basiliensis Commentariorum libri II* (1440), new edition and translation by Denys Hay and W. K. Smith in Oxford Medieval Texts (1967).

an amusing story teller;[1] and he did more than anyone else to spread in Central Europe an interest in Latin eloquence, in classical education,[2] and in history and geography.[3] As an intermediary he excelled all his contemporaries, and for that reason he deserves a place in the history of scholarship. A year after his election to the papal throne he signed the document of the foundation of the university of Basle, which in contrast to the old universities at once became the very home of humanism and classical studies that it has remained to the present day.[4] But then, to the disappointment of hopeful humanists and scholars, he lost interest in letters and after a fruitless attempt to convert the sultan who had conquered Constantinople concentrated his energy for the rest of his days on the organization of a crusade against the Turks. Some of his own great works on history and geography which contained large excerpts from ancient sources remained fragmentary, but they exercised an enormous influence.[5] Even the explorer Christopher Columbus,[6] writing from Jamaica in 1502/3, quoted his description of the Massagetes. Enea Silvio's *Historia Bohemica* and his *Germania* became standard works, testifying the continuous connection of Italian humanism with Central Europe.

The activity of Italian scholars was extended also to Western Europe, to France and England. Petrarch and other exiled Italians had found a new home in the south of France, where they had the benefit of easier access to the libraries of French cathedrals and monasteries.[7] But how far did France benefit from the presence of Italian visitors or immigrants? During his sojourn in Paris in 1361 Petrarch had a faithful companion in the priest Pierre Bersuire,[8] who started to translate Livy for King John the Good; an interest in

---

[1] *Historia de duobus amantibus* (1444), for which he freely used the story of a love-adventure of the Chancellor Caspar Schlick in Vienna (reprinted Budapest 1904).

[2] *De eruditione puerorum* (1450) addressed to the young king of Bohemia and Hungary, Ladislav (reprinted Washington 1940).

[3] On his writings on contemporary history see above, p. 59, nn. 3 and 4.

[4] August Rüegg, *Die beiden Blütezeiten des Basler Humanismus*. Eine Gedenkschrift zur Fünfhundertjahrfeier der Basler Universität (1960); cf. Guido Kisch, 'Forschungen zur Geschichte des Humanismus in Basel', *Archiv für Kulturgeschichte* 40 (1958) 194 ff.

[5] *Historia utique gestarum locorumque descriptio*, less exactly called *Cosmographia*; on contemporary history the autobiographical *Commentaria rerum memorabilium*.

[6] 'De Asia' c. 12 = *Opera* (1571) p. 289, see B. Widmer (above p. 59 n. 3) p. 386.2; a copy of the first part of the *Historia* was in the possession of Columbus: (Henry Harrisse), *Don Fernando Colon, Historiador de su padre, Ensayo critico* (Sevilla 1871) pp. 67, 75. On Columbus and his considerable knowledge of ancient and humanistic literature see A. Cioranescu, *Colón, humanista* (Madrid 1967).

[7] See above, p. 12.

[8] F. Simone, 'Il Rinascimento Francese', *Biblioteca di Studi Francesi* 1 (1961) 23.

Roman historians and the effort to translate important texts have thus been characteristic of France from the beginnings. Jean de Montreuil (1354–1418)[1] and his circle admired Petrarch and Salutati, and tried to imitate the new Latin style. But he and his friends remained an isolated group; comparing similar episodes in Italy and Germany, we may speak of a French prehumanism.

In the fifteenth century the scholastic philosophy was still very powerful with a special inclination towards encyclopedism; but the chancellor of Paris university, Jean Charlier de Gerson[2] (d. 1429), and a few others showed an increasing familiarity with the classical texts. The royal court, residing at different places, became another centre of cultural life, open to influence from abroad, especially from Italy. An ambitious cultural nationalism grew up there which tried to rival Italian humanism, but was not itself very fruitful in scholarship. In contrast to this nationalistic line, a more universal spirit was represented in Paris by Lefèvre d'Étaples (Faber Stapulensis)[3] under the spell of the Florentine Neoplatonism, which also greatly influenced Margaret of Navarre and her literary court.[4] It was from Italy that Greek scholars like Janus Lascaris found their way to France and spread a better knowledge of Greek language and literature.[5]

There is no evidence that Italian humanists came as missionaries to the Iberian peninsula. Of course, Poggio's[6] world-wide correspondence extended also to the far west, but not his travels. On the other side, Spaniards like King Alfonso of Aragon, before he went to Naples, and the Castilian court took some notice of the new movement, and finally young scholars began to visit its places of origin.[7]

The British isles[8] were reached by Italian humanists later than most of the continental countries. Poggio[9] accepted in 1418 an invitation from the bishop of Winchester, Cardinal Henry Beaufort, whom he had met at the Council of Constance. Though this visit was disappointing

---

[1] G. Billanovich et G. Ouy, 'La Première Correspondance échangée entre Jean de Montreuil et Coluccio Salutati', *IMU* 7 (1964) 337 ff. A. Combes, 'Jean de Montreuil et le chancelier Gerson', *Études de philosophie médiévale* 32 (1942), mostly about problems of the letters and theological questions.

[2] See above, n. 1; Ioannis Carlerii de Gerson, *De mystica theologia* ed. A. Combes, 1958 (Thesaurus mundi), critical edition with bibliographical references.

[3] On his relation to Erasmus see below, p. 100.

[4] A. Tilley, *The Literature of the French Renaissance* I (1904) 103.

[5] Cf. above, p. 48 and below, pp. 65 and 94 ff.

[6] Walser, *Poggius* p. 298; cf. Voigt II 356 f.

[7] See below, p. 94.

[8] R. Weiss, *Humanism in England during the Fifteenth Century* (2nd ed. 1957).

[9] See above, p. 33.

for Poggio, at least it gave English clergymen what seems to have been their first contact with a prominent Italian scholar. Otherwise they met the papal officials and collectors who came over from Italy year after year. Of these the best known is the 'nuntius et collector' Piero de Monte, a pupil of Guarino, an indefatigable letter-writer[1] and the author of probably the first humanistic treatise in England on virtues and vices. Members of the British nobility and clergy became patrons of the new learning, and gave hospitality to foreign scholars. Humphrey, Duke of Gloucester, a son of Henry IV, had several Italian humanists,[2] Piero de Monte, Tito Livio Frulovisi, and others, at his service and collected manuscripts from Italy for his library which he afterwards left to the university of Oxford, but he is supposed to have read the classics in French translations. A number of Italian humanists, Bruni, Pier Candido Decembrio, Castiglioni, dedicated their works to him, as to the Pope or to the Medici. Enea Silvio came to England and Scotland,[3] as an envoy from the Church Council at Basle;[4] and even the Greek grammarian and teacher Manuel Chrysoloras,[5] whom Salutati had brought to Florence, paid a visit to this country.

The contact, however, between Italy and the north was kept up also from the other side, from north to south. There were first of all German students who in accordance with a long tradition crossed the Alps to study Roman and Canon law at the old university of Bologna; but now, attracted by the new learning, many went on to other places, to Padua, Venice, Florence, Rome. They mostly came from old or rich families in the free cities, as for instance Willibald Pirckheimer[6] from Nuremberg and Conrad Peutinger from Augsburg.[7] Peutinger always felt a deep obligation towards his Italian teachers; even forty years after his stay in Italy (1482–8) he printed a collection of excerpts from their lectures. In Florence he was lucky enough to meet Politian and Pico, and he tried hard, but in vain, to get the famous manuscript of the

---

[1] 'Piero de Monte. Ein Gelehrter und päpstlicher Beamter des 15. Jahrhunderts. Seine Briefsammlung'. Herausgegeben und erläutert von Johannes Haller. *Bibliothek des Deutschen Historischen Instituts in Rom*, vol. 19 (1941).

[2] *Opera* T. Livii de Frulovisiis, rec. C. W. Previté-Orton (1932); see especially pp. xiii f. and xxxv f.

[3] G. Paparelli, *Enea Silvio Piccolomini* (1950) pp. 51 ff.

[4] See above, p. 59.

[5] See above, p. 27.

[6] Willibald Pirckheimer, *Briefwechsel*, hrsg. von E. Reicke und A. Reimann 1 (1940, reprinted 1970), 11 (1956). H. Rupprich, 'W. Pirckheimer, Beiträge zu einer Wesenserfassung'. *Schweizer Beiträge zur allgemeinen Geschichte* 15 (1957) 64 ff.

[7] 'Conrad Peutinger und die humanistische Welt', *Augusta* (1955) 179–86 = *Ausgewählte Schriften* (1960) pp. 222 ff., and 'Augsburger Humanisten und Philologen', *Gymnasium* 71 (1964) 190 ff.

Digests for the Emperor Maximilian. He was imbued for life by the Neoplatonic spirit of the Florentine Academy; and Pomponio Leto's Academy in Rome inspired him to collect the Roman antiquities of his own country and to found a 'Sodalitas literaria Augustana'. He remained in correspondence with his Italian friends when he became the 'cancellarius' of his native city, highly esteemed by Erasmus for 'gravitas' and 'prudentia'.

Another group of German students were itinerant poets, a sort of caricature of the Italian wandering humanists. One of those adventurous travellers through many lands was Peter Luder (*c.* 1415–74), who spent some time in Ferrara as a pupil of the younger Guarino, before going on to Pavia and Rome and even to Greece, not without some difficulty. Called to Heidelberg[1] in 1456, he said in the announcement of his lectures: 'studia humanitatis i.e. poetarum, oratorum ac historiographorum libros publice legi instituit';[2] Salutati's new formula,[3] we see, was already accepted outside Italy.

In the same category, but of a higher order was Conrad Celtis (1459–1508),[4] who wrote the best Latin poems on this side of the Alps (*Amores*, 1502, cf. *Ars versificandi*, 1486) and was the first German to be crowned as a poet by the Emperor. After studying in Cologne and lecturing on Platonic philosophy at several German universities, he journeyed to Italy where he tried in the short space of six months to learn from Battista Guarino in Ferrara, from Musurus in Padua, and from Pomponio Leto in Rome. Celtis, though able, always remained the pupil of his Italian masters, and it is hardly correct to label him as 'the German arch-humanist'. In one respect, however, he seems to have been inventive; in 1497, when he was lecturing on Horace in the university of Ingolstadt, he[5] induced the composer Petrus Tritonius to

---

[1] Peter Luder was not an isolated figure, see G. Ritter, 'Petrus Antonius Finariensis, der Nachfolger Peter Luders in Heidelberg. Ein Beitrag zur Geschichte des Frühhumanismus am Oberrhein', *Archiv für Kulturgeschichte* 26 (1936) 89 ff.

[2] E. König, 'Studia humanitatis und verwandte Ausdrücke bei den deutschen Frühhumanisten', *Beiträge zur Geschichte der Renaissance und Reformation* (Festgabe für J. Schlecht) (1917) p. 203.

[3] See above, p. 25.

[4] Conrad Celtis, *Opuscula* ed. K. Adel (1966); *Briefwechsel* hg. u. erläutert von H. Rupprich, (1934); *Amores* ed. F. Pindter (1934).

[5] This is clearly attested in the preface of Simon Minervius to the *Varia carminum genera* of L. Senfl (1534), p. 4a (Tritonius) 'qui . . . ductu et auspiciis Conradi Celtis . . . hortatu praeceptoris . . . harmonias composuit'; Tritonius's compositions were printed in 1507. Notable contemporary composers took up this metrical style for a short time in the early sixteenth century. I am not sure whether Athanasius Kircher's composition of a melody to Pindar's first *Pythian* (*Musurgia* 1 [1650], 541) was still in this tradition as J. Müller-Blattau, *Herm.* 70 (1935) 103 ff. tried to demonstrate. On the decisive part played by Celtis see

make tunes for nineteen odes to be sung daily at the end of the lectures *secundum naturas et tempora syllabarum et pedum.* Otherwise he was an imitator of the Italian humanists, Germanizing their manners and ideas; his earnest desire was that his country should not remain behind Italy. The Italian humanists were proud of their Roman ancestry and tried to renew ancient culture; by analogy, Celtis was the first to emphasize the value of German antiquity. He naturally began with an edition of Tacitus' *Germania* (1500), the work to which Nicholas of Cusa and Enea Silvio had paid so much attention. He also searched for manuscripts, not of ancient, but of medieval authors of German origin; in 1492/3 he actually discovered the Latin plays of the medieval German nun Hrosvitha (tenth century) in the monastery library of St. Emmeram in Regensburg, and a Latin epic poem *Ligurinus* on the Emperor Frederick I (twelfth century). It was, therefore, a classical scholar, working after the Italian fashion, who initiated German scholarship.[1]

Celtis had the good luck to find a medieval copy of an ancient map, which he left to his friend Peutinger for publication. Though it was not published by him, but only at the end of the sixteenth century by Marcus Welser, it is still called *Tabula Peutingeriana.* Celtis conceived the plan of a *Germania illustrata*, on the lines of Flavio Biondo's *Italia illustrata*, about the origin and past of the German people (as opposed to the empire). He also organized literary societies, modelled on Pomponio Leto's, all over Central Europe. Celtis's humanistic nationalism did no harm in itself; but it unfortunately could lead to Teutonic mummery, to Celtis's own fantastic confusion of the original wisdom of the Druids with Neoplatonic mysticism, and to Hutten's identification of German virtue, as praised by Tacitus, with early Christian piety.

Much closer links between Italy and the north than those made by itinerant student and knights-errant were formed by Nicholas of Cusa[2] and Rodolphus Agricola, who studied in Pavia and Ferrara with two interruptions from 1468 to 1479; but we have already mentioned Nicholas's Italian visits, and we shall have to deal with Agricola as one of Erasmus's predecessors.

Zeitschrift für deutsche Philologie 46 (1914/15) 287 f. and *Gymnasium* 71 (1964) 198 f.; W. Salmen, *New Oxford History of Music* III (1960) 370 f. 'the didactic metrical ode: the incentive came from the cultured aristocracy', neglected the evidence about Celtis in the preface to Senfl's *Carmina* quoted above.

[1] J. Dünninger, 'Geschichte der deutschen Philologie', *Deutsche Philologie im Aufriss* I² (1957) 87 ff., laid the right stress on the importance of humanism and Reformation for the origin of German scholarship in general, but perhaps not enough on Celtis's personal merits.

[2] See above, p. 39, and below, pp. 69 and 139.

From France Jean de Montreuil,[1] Petrarch's great admirer, was sent by King Charles VI in 1412 to Rome and Florence where he met Bruni and Niccoli and brought back copies of Plautus, Varro, and Livy. In the second half of the fifteenth century Robert Gaguin (1433–1501) went from Paris on various successful missions to Italy and Germany, culminating in his appointment as royal ambassador to these countries and to England. He had a genuine interest in the new learning and produced a number of Latin works from *De arte metrificandi* (1478) to his history of France, *De origine et gestis Francorum compendium* (1495). It was mainly this history that earned him his reputation as the leading Paris humanist. The young Erasmus submitted a draft of his *Antibarbari* to Gaguin's notice, anxious to learn what he thought of it, and Gaguin's letters are full of references and allusions to classical texts.[2] Early in the sixteenth century the foremost French classical scholar, Guillaume Budé, was charged with diplomatic missions to the Medicean popes, Julius II and Leo X. It was the royal court which continued to further France's cultural connection with Italy by using distinguished learned persons as delegates.

The first young visitors from the Iberian peninsula seem to have been the Portuguese Ayres Barbosa, who was lucky enough to find the best teacher in Florence, Angelo Poliziano, and then the Spaniard Hernán Núñez. They both taught Latin and even Greek in the university of Alcalá. A Cretan, Demetrios Dukas,[3] moved to Spain from Venice where he had worked for the Aldine Press. Both Núñez and Dukas took part in preparing the most celebrated Spanish publication of the early sixteenth century, the Polyglot Bible, supported from 1502 onwards by the great cardinal Francisco Ximenes de Cisneros (1437–1517).[4] The Greek of the New Testament was completed and printed at Alcalá in January 1514, and that of the Old Testament in 1517, but it was not before 1521/2 that a relatively small edition of 600 copies was published.[5] It was called 'Complutensis' from Complutum, the Latin name of the town Alcalà. Antonio de Lebrija[6] (1444–1522)

---

[1] Cf. above, p. 61.

[2] On Gaguin's activity see P. S. Allen's notes, Erasm. *Ep.* 43; on Gaguin's criticism of the *Antibarbari* see 'Die Wandlungen der Antibarbari' in *Gedenkschrift zum 400. Todestage des Erasmus* (1936) p. 54 = *Ausgewählte Schriften* (1960) p. 192, where I have emended *Ep.* 46.41 'si in formose Veneris cute nervos curem' into 'nevos'. Allen had not noticed the allusion to Hor. *Sat.* 16.67.

[3] D. J. Geanakoplos, *Greek Scholars in Venice* (1962) pp. 223–55.

[4] M. Bataillon (below p. 95 n. 4, pp. 1–78 and *passim*).

[5] F. J. Norton, *Printing in Spain 1501–20* (1966) pp. 38 ff.

[6] Bataillon (below, p. 95 n. 4) pp. 24–42. F. G. Olmedo, *Nebrija* (Madrid 1942); cf. Geanakoplos (above n. 3) pp. 273 f.

had studied Hebrew as well as the classics in Italy for twenty years. After 1502 he was recalled to his native country by Cardinal Ximenes; for as a 'trilinguis', he was greatly needed to help with the Polyglot Bible. As its publication was delayed by various circumstances, Froben in Basle hastened to forestall it; issuing in 1516 3,300 copies of Erasmus's Greek and Latin text of the New Testament, he won the field.

In England we find several gifted students who completed their education in the schools of famous Italian humanists, especially in Guarino's school in Ferrara. In the first half of the Quattrocento the Hundred Years' War prevented English students from going to Paris, and by the end of the century we can notice some results of their visits to Italy. The first name of importance is that of Thomas Linacre (1460–1524)[1] who was taught by his uncle, the Benedictine Prior William Sellyng, in his birth-place Canterbury and then brought by him to Rome and Florence, where he became a pupil of Politian and Chalcondyles, together with the future pope Leo X. He remained in Italy from 1487 to 1498, continuing his studies in Rome with Pomponio Leto and later in Padua and Venice, where he made friends with Ermolao Barbaro and with Aldus Manutius of whose 'academy' he became a member. After his return to London, he gave lectures which were attended by Thomas More. Two friends studied with Linacre in Italy: William Grocyn and William Latimer. Grocyn may have been the first to teach Greek in an English university, probably at Exeter College, Oxford. No upheaval was caused by recent developments nor even a break with the past; collecting books, teaching in schools and colleges, translating ancient texts into English, writing prose and verse in Latin and later also in Greek, went on steadily and quietly, shifting of course to a more classical style. No really characteristic feature of English humanism and scholarship is to be discerned before the time of John Colet.

If there had been only the Italian visitors to the north or the eager students and knights-errant from the north travelling through Italy, the role of the transalpine countries in the history of scholarship would be rather modest. All these contacts were in fact invaluable in preparing the way for northern humanism, but its main stream was to derive from another, native source.

---

[1] R. Weiss, 'Un allievo Inglese del Poliziano: Thomas Linacre' in 'Il Poliziano e il suo tempo', *Atti del IV Convegno internazionale di Studi sul Rinascimento* (1957) 231–6.

PART TWO

HUMANISM AND SCHOLARSHIP IN THE NETHERLANDS AND IN GERMANY

VI

DEVOTIO MODERNA

IN the north there was no great poet able to inspire his contemporaries
with a new love of the classics. But a new religious movement of laymen,
who called themselves *fratres communis vitae*, 'brethren of common life',
tried to find their way to the original texts of the Bible and the Church
Fathers, especially St. Jerome; and through these ecclesiastical writings
they came to ancient Greek and Roman literature. This movement did
not depend on the Church or on the Scholastic tradition, but tended to
a new form of piety, simpler and more individual, from which its usual
name *Devotio Moderna* was derived.[1] It was founded in the second half of
the fourteenth century by Geert Groote (Gerhardus Magnus) in the
city of Deventer on the river Yssel in Holland. From this city, which
always remained its centre, the movement spread eastwards through
northern Germany to the Vistula, founding schools everywhere, which
afterwards had their own printing offices.

The constitution of the communities insisted on *pure* texts, 'lest one's
conscience might be hurt by some improper version'; medieval Latin
was abhorred as 'barbarous', and the classics enjoyed a pre-eminence in
education. Nicholas of Cusa[2] and Thomas à Kempis were educated in a
Deventer school. The headmaster of Erasmus's school at Deventer,
Alexander Hegius, even wrote a poem 'on the utility of the Greek
language'. Hegius's friend, the philosopher and humanist Rudolf

---

[1] The term was consciously used by the founder Geert Groote, see M. Ditsche, 'Zur Herkunft
und Bedeutung des Begriffes Devotio Moderna', *Historisches Jahrbuch* 79 (1960) 124 ff.—
On the history of the Devotio Moderna see A. Hyma, *The Christian Renaissance* (1924, 2nd ed.
1965) and *The Brethren of the Common Life* (1950); see also *The Youth of Erasme* (1930, 2nd ed.,
enlarged 1968; cf. below, p. 80 n. 1) pp. 21 ff., 88 ff. and R. R. Post, *Modern Devotion.
Confrontation with Reformation and Humanism* (Leiden 1968). The essential importance of
Devotio Moderna for humanism in the north was stressed before by A. Roersch, *L'Humanisme
belge à l'époque de la renaissance* (1910) pp. 9 ff. and P. Mestwerdt, *Die Anfänge des Erasmus,
Humanismus und Devotio Moderna* (1917).—I am, of course, aware of recent tendencies to
belittle the influence of Devotio Moderna. But having returned from time to time to the
preserved texts of the whole circle and to the early letters of Erasmus, who spent twelve years
with the brethren, I see no reason to change my mind. The most radical devaluation I know
is contained in the paper of H. M. Klinkenberg, 'Devotio moderna', of which a summary is
printed in *Jahres- und Tagungsbericht der Görres-Gesellschaft* (1957) pp. 43 f.

[2] On Nicholas of Cusa and his relation to Italian humanism see above, p. 39.

Agricola,[1] an admirer of Petrarch and a Neoplatonist in the Florentine style, seems to have coined the term *Philosophia Christi*[2] to describe his teaching, the object of which was to mediate between ancient wisdom and Christian faith, always aiming at the improvement of Christian piety.

[1] Cf. Agricola, 'De formando studio lucubrationes', *Opera* II (1539) 193 ff. and especially Mestwerdt (above, p. 69 n. 1) p. 162.

[2] Cf. Clem. Al. *Strom.* VI 8.67.1 (II p. 465.21 St.) τῆς κατὰ Χριστὸυ φιλοσοφίας.

VII

ERASMUS OF ROTTERDAM

ERASMUS was a pupil of the brethren of common life,[1] and the term *Philosophia Christi* which he took from Agricola[2] brings us to the centre of his Christian humanism. Indeed to understand his philology we must start from this conception; for it was the inspiration of his scholarship and of his enormous services to learning.[3]

Erasmus was born at Rotterdam, and the evidence seems to favour 1469, not 1466, as the year of his birth.[4] He was sent in 1478 to the school at Deventer, entered the Augustinian monastery at Steyn near Gouda in 1487, and was ordained priest in 1492. In the following year he entered the service of the bishop of Cambrai and left his monastery for good. In Paris Erasmus came into contact with the late Scholastic philosophy,[5] and continued his classical studies, concentrating on the

---

[1] *Opera omnia* recognovit Iohannes Clericus, 10 vols. (Leyden 1703–6, reprinted 1961/2). *Opus epistularum* (abbrev. *Ep.*) edd. P. S. Allen, H. M. Allen, H. W. Garrod, 12 vols. (Oxford 1906–58). A new French edition has been in preparation since 1967: *La Correspondence d'Érasme*, tradition intégrale sous la direction d' A. Gerlo et P. Foriers, to be in 12 vols., 1: 1484–1514. *Opuscula*, a supplement to the *Opera omnia* ed. W. K. Ferguson. A new complete annotated edition in 30 vols. is being published by the North Holland Publishing Company, Amsterdam: Erasmus, *Opera omnia* (1969 ff.) edited and annotated by K. Kumaniecki, R. A. B. Mynors, and others. Selected works and monographs are quoted below at the appropriate places. For an extensive bibliography up to 1966 see Kohls (below, n. 3) II 137 ff.

[2] The often repeated formula 'studii forma' in Erasmus's *Enchiridion militis Christiani* seems also to have been derived from Agricola.—On the relations between Agricola and Erasmus see P. S. Allen, Erasm. *Ep.* I 106 f.

[3] I prefer 'philosophia Christi' (or 'Christiana') to the more usual term 'theology'. The recently published two-volume *Die Theologie des Erasmus* by E. W. Kohls, Theologische Zeitschrift, Sonderband 1, 1 and 2 (Basel 1966) is to be welcomed, although the author, after a penetrating critical survey of previous research on the same subject, limits his own very careful analysis to the early writings of Erasmus. This may be justified by the fact that they contain all his fundamental ideas and these ideas changed remarkably little in the course of his life.

[4] 'Die Wandlungen der *Antibarbari*', *Gedenkschrift zum 400. Todestag des Erasmus von Rotterdam* (1936) p. 53 = *Ausgewählte Schriften* (1960) p. 191, see also below, p. 80 n. 1; unfortunately there is a misprint n. 11: 'quinquaginta' instead of 'quadraginta'. In a passage so far not used in the discussion *Ep.* 2136 (31 Mar. 1529) he called himself 'sexagenarius', which would be correct only if he was born in 1469. E. W. Kohls, 'Das Geburtsjahr des Erasmus', *Theologische Zeitschrift* 22 (1966) 96 ff., 347 ff. pleads most strongly for the earlier date (1466) without convincing me.

[5] In his pungent style he compared the traditionalism of the Sorbonne with the dried skin of Epimenides of old. Cf. the informative and judicious dissertation of C. Dolfen, *Die Stellung des Erasmus zur scholastischen Methode* (Diss. Münster 1936).

study of the Greek language, which he could not acquire so well anywhere else.[1] In 1499 one of his English pupils, Lord Mountjoy, invited him to England; there while living in Oxford at New College, he met John Colet[2] of Magdalen. This meeting was decisive for his life's work, the study of the Bible and of the Church Fathers.

Colet had lived for some time in Italy without becoming an 'Inglese italianato'; but he was consciously indebted to the Florentine Platonism.[3] His lectures on the Epistles of St. Paul to the Romans and to the Corinthians, delivered in the three years between his return from Italy and his meeting with Erasmus, contained express references to t Theologia Platonica of Ficino. We saw that the Florentine way reconciling antiquity and Christianity- was by some kind of myst symbolism. But Colet's way—and, as we shall see, Erasmus's way—w² entirely different; Colet's way of solving the problem was, I should say a truly philological one. He neither applied the allegorizing method o the Middle Ages which still flourished in England in the fifteentl century nor took over the symbolic method of the modern Platonists. He tried to comprehend the religious feeling of St. Paul as a real person who wrote real letters, and to understand the meaning of the epistles, their sentences and words, in detail. In this he was obviously following the example of Petrarch's interpretation of Cicero's letters.[4] The aim of Colet's interpretation was not to discover new subtleties of rare knowledge—he rather sweepingly condemned Scholasticism—but to bring about the 'restitution' (this is his own word, Erasm. *Ep.* 108. 58) of true piety in practical life. So in principle he agreed with the tendencies of the Devotio Moderna, with which Erasmus had been familiar since his school-days.

Colet's ideas did not effect an immediate transformation in Erasmus's life; but they cleared the air and became a light to guide him through dark years of incessant labours to his greatest achievements. Above all, they strengthened his belief that Christian piety could be revived through the spirit of ancient *humanitas*; at least the outlines of this programme were visible in an early version of the *Antibarbari* which

---

[1] The early correspondence of Erasmus is a priceless source for our knowledge of teachers of Greek in Paris, see below, p. 102.

[2] J. Colet, *Opera* ed. J. H. Lupton, 5 vols. (1867–76, repr. 1965/6). J. H. Lupton, *A Life of John Colet* (1887, repr. 1961). L. Borinski, *Englischer Humanismus und deutsche Reformation* (1969) pp. 12 ff. On his Lectures see P. Albert Duhamel, 'The Oxford Lectures of J. Colet', *Journal of the History of Ideas* 14 (1953) 493 ff. For new manuscript material, partly autograph, from All Souls College, Oxford, and for an exhaustive bibliography see Sears Jayne, *John Colet and Marsilio Ficino* (1963) especially pp. 77 f.

[3] See above, pp. 57 ff.   [4] See above, pp. 9 f.

pleased Gaguin in Paris and then impressed Colet in England.[1] One embarrassing difficulty for Colet was that he had not learned Greek, the importance of which for educational purposes had been insisted on by Battista Guarino as long ago as 1459.[2] 'Without it we are nothing' was Colet's own confession to his young friend, who repeated it and was followed by many others, J. J. Scaliger finally declaring: 'Who knows no Greek knows nothing.'[3] Now it was urgently necessary for Erasmus to polish his rather poor Greek, and for this reason he had to go back to Paris. But he returned to England for two important visits: in 1505/6 to London, where Colet was dean of St. Paul's and Thomas More already a famous lawyer, and from 1509 till 1514 partly to London and partly to Cambridge,[4] where he lived in 1512/13 in Queens' College teaching some Greek and preparing his Greek New Testament and Jerome's Latin text. Between his two visits to England he made use o an opportunity of going to Italy, and in 1508 became a member of the Aldine Academy in Venice;[5] his time there was spent entirely in libraries, printing offices, and learned societies. Afterwards he travelled frequently up and down the Rhine between Louvain[6] and Basle. There having found his great printer, John Froben, he settled down from 1521 to 1528;[7] he was forced by the troubles of the Reformation in Basle to withdraw in 1529 to the Catholic city of Freiburg im Breisgau, but he returned to his beloved Basle in 1535 and died there in 1536.

This course of life divided between various countries and nations was characteristic of Erasmus. In one of the most famous passages of his letters he told Zwingli (*Ep.* 1314.2 ff.): 'Ego mundi civis esse cupio communis omnium vel peregrinus magis.'[8] But equally characteristic

---

[1] *Ausgewählte Schriften* p. 193.   [2] See above, p. 53.

[3] *Epist.* (editio Lugd. Batav. 1627) p. 51.

[4] *Erasmus and Cambridge, The Cambridge Letters of Erasmus* transl. by D. F. S. Thomson, introd., comm. and notes by H. C. Porter (1963).

[5] Cf. above, p. 56. In general see P. de Nolhac, *Erasme en Italie* (2nd ed. 1898); more comprehensive, as the title indicates, is the book by A. Renaudet, *Érasme et l'Italie*, Travaux d'humanisme et renaissance 15 (1954). After his visit to Venice the pages of the Aldine edition of the *Adagia* were numbered, and others followed this example.

[6] Henry de Vocht, *History of the Collegium trilingue*, 4 vols. (1951–5) see esp. III 406.

[7] No doubt the years at Basle should be regarded as the most important of his life.

[8] In quoting these words one should never forget the very important sentence that follows: 'Utinam contingat asscribi civitati coelesti; nam eo tendo.' His 'Stoic' cosmopolitanism had a Christian end. Even C. R. Thompson, 'Erasmus as Internationalist and Cosmopolitan', *Archiv für Reformationsgeschichte* 46 (1955) 167 ff., left out the reference to the 'celestial city'. With due reserve I suggest that 'asscribi civitati coelesti; nam eo tendo' was derived from one of the Church Fathers whom Erasmus knew so well; cf. Greg. Naz. *Orat. in laudem sororis Gorgoniae* c. 6 (*PG* 35. 796). Γοργονίᾳ πατρὶς μὲν ἡ ἄνω Ἱερουσαλήμ . . . πόλις ἐν ᾗ πολιτευόμεθα καὶ πρὸς ἢν ἐπειγόμεθα (in quam cives asscripti sumus et ad quam tendimus). It is likely that similar expressions occur elsewhere in Patristic literature, Greek and Latin. E. Arnold suggests that Paul's Epistle to the Hebrews 12:22 might be the primary source; cf. also 13:14.

and important was the continuity of Erasmus's thought. From the beginning to the end he remained faithful to the fundamental conception of 'philosophia Christi'. This 'philosophy' was inseparably connected with the Socratic theory that knowledge is the necessary condition of acting well and that ignorance leads to evil. The struggle against the ignorance of his age was the struggle against evil. The spiritual decline, so he felt, was most clearly revealed in the deterioration of language. And so it was with language that spiritual and moral renaissance must begin. To Erasmus the revision of the ancient texts through an improved knowledge of the classical languages was the highest task of all. This revision encompassed the very sources of spiritual and moral life, first the scriptures, then the classics, and finally the Church Fathers, who formed the connection between the classics and the Bible. By purifying these texts of the errors of centuries and restoring them to their obviously simple truth, it would be possible to check the corruption of his own time. For this not only would new editions be necessary, but particularly commentaries,[1] paraphrases,[2] and translations.

Erasmus outlined his method of interpretation in the introduction, on 'Methodus', to his first edition of the New Testament (1516), which he enlarged in the second and third editions of 1519 and 1522 and later published separately as a much longer essay with a changed text and title: *Ratio seu methodus compendio perveniendi ad veram theologiam*.[3] This method was not a narrow verbal criticism, as has often been alleged by theologians who had too little acquaintance with the text of his writings; on the contrary, according to his rules, in order to understand the details, we must know and understand the whole of the divine word: 'audi sermonem divinum,[4] sed totum audi'. All particulars must be shown to be in harmony with the living body of Christian doctrine which is both gospel and tradition: 'sensus respondeat ad *orbem* illum doctrinae Christianae.' In his constant appeals to reason he was free from overbearing rationalism; unlike John Colet he did full justice to the great medieval philosophers such as St. Thomas Aquinas, but

[1] See above, pp. 54 f. on commentaries in the Italian Renaissance.

[2] On ancient paraphrases see *History* [1] 219.5 Addenda.

[3] Neudruck und Bibliographie in Erasmus, 'Ausgewählte Werke', hg. von H. Holborn, *Veröffentlichungen der Kommission zur Erforschung der Geschichte der Reformation und Gegenreformation*, München 1933, pp. 150–62 'Methodus', pp. 175–305 'Ratio'. See my review in *Gnomon* 12 (1936) 625 ff.

[4] Erasmus prefers 'sermo' to 'verbum' (see especially his translation of John 1 : 1). Th. More, *Correspondence* ed. E. F. Rogers (1947) ep. 83 p. 179, passionately supports Erasmus's rendering of λόγος. On the great importance of the term 'Verbum divinum' see Max Schoch, *Verbi Divini Ministerium*, vol. 1 Verbum—Sprache und Wirklichkeit, 1968; this volume deals only with the Reformers and with Luther.

rejected the logical subtleties of the later degenerate theology, calling attention to the limits of human knowledge and research.

All histories of classical scholarship are accustomed to quote a saying of Gottfried Hermann: 'est quaedam nesciendi ars et scientia', and Wilamowitz[1] once remarked that already Hugo Grotius had said the same: 'nescire quaedam magna pars sapientiae est.' But, even before that, Erasmus had written in a letter[2] often printed and certainly known to Grotius: 'et scientiae pars est quaedam nescire.' What this expresses is not a trivial scepticism, but the natural shyness of religious men, their fear of transgressing the limits of human reason. 'Praestat venerari quaedam quam scrutari', he said in another place,[3] and, oddly enough, Goethe used almost the same words without knowing the Erasmian passage.[4] There is an extremely strong feeling of the suprarational in Erasmus. But he believed that within its limits the human mind is obliged to make the greatest possible effort to do its duty: 'cognoscere, intellegere, scire'. The editor and interpreter as well as the listener and reader are expected to use the severest critical judgement and to be on their guard against easy submission to authority.

Knowledge can never be dangerous to true religion, as Erasmus continuously replied to innumerable attacks; on the contrary, the danger lies in ignorance as in a bad text or a false interpretation. There are in the various prefaces to the New Testament quite a number of ironical retorts against people who belittled his grammatical efforts and wanted the traditional doubtful or wrong readings to be retained. 'Well,' he said, 'God may not be offended by solecisms, but he is not delighted by them either'; and a little more seriously he asked: 'Why trouble about punctuation? But a wrong hypostigma, a comma, such a small thing produces a heresy (tantula res gignit hereticum sensum)', and he added examples. At the end of the *Methodus*, declaring in a tone of grave solemnity 'Abunde magnus doctor est qui pure docet Christum', he made abundantly clear his belief that there can be no way to the *pure* sources of the evangelical truth, to 'veritas evangelica', without scholarly criticism.[5]

Scholarly criticism was one of the glories of the Hellenistic age. Among the Italian scholars it was Lorenzo Valla who first showed a truly critical spirit, as we have seen.[6] We must not forget that Erasmus said of some works of the so-called Epistolographi, going under the names of Brutus, Phalaris, Seneca, and Paulus that they would better

<hr>

[1] *Geschichte der Philologie* p. 49.   [2] *Ep.* 337. 419.   [3] *Op.* IX 273 B.
[4] *Ausgewählte Schriften* p. 246.   [5] See *Ausgewählte Schriften* p. 214.   [6] See above, pp. 38 ff.

have been termed 'declamatiunculae' than 'epistulae'.[1] This is a small step beyond Valla on the way to Bentley.

Hellenistic scholarship was centred on Homer.[2] In the course of this chapter we have been reminded of many points from Hellenistic scholarship. Now, however, the centre was no longer Homer but the New Testament. As it had been said of Homer that he was his own interpreter, so the 'scriptura sacra' was called 'sui ipsius interpres'. Erasmus's only worthy predecessor in the field of biblical criticism and exegesis was Lorenzo Valla,[3] to whom he felt himself deeply indebted. The correct starting-point for assessing Erasmus's several services to learning must always be his edition of the Greek text of the New Testament, published in 1516 in Basle.[4] The genius of Holbein[5] created in the following years the pictures of Erasmus in which the harmony of the humanist and the evangelist is completed. Erasmus, however, himself added to his portrait on the medal by Quentin Metsys the inscription: τὴν κρείττω τὰ συγγράμματα δείξει, 'the writings will show the better picture'.[6]

Erasmus's first care was for good manuscripts; though the prefaces of this time say nothing on the subject, there are occasional references in the letters.[7] We learn that he had 'some' Greek manuscripts in Cambridge and hoped to find better ones in Basle, but was disappointed. He prepared a manuscript of the fifteenth century for the printer, and gave another manuscript of the twelfth century, borrowed from Reuchlin, to the corrector, then one of the thirteenth century which he thought to be more accurate, and finally two other, fifteenth-century manuscripts. For the later editions of 1519, 1522, 1527, and 1535, he

---

[1] See T. O. Achelis, 'Erasmus über die griechischen Briefe des Brutus', *Rh. Mus.* 72. (1917/18) 633 ff. with the respective references to the whole problem (Dionys. Areop.). See also below, p. 152.

[2] See *History* [1].105 ff.

[3] See above, pp. 38 ff.

[4] With the title *Novum Instrumentum*; the later editions returned to the traditional title. On the introductory essays see above, p. 74.

[5] Hans Diepolder, 'Hans Holbein d.J., Bildnisse des Erasmus von Rotterdam', *Der Kunstbrief*, Heft 56 (1949).

[6] Cf. Ov. *Tr.* 1 7.7 ff. 'effigiemque meam . . . sed carmina maior imago', Mart. IX 76.9 ff. 'pictura . . . in chartis maior imago meis'; see W. Speyer, 'Naucellius und sein Kreis. Studien zu den Epigrammata Bobiensia', *Zetemata* 21 (1959) 59. The Greek form seems to be Erasmus's invention.

[7] *Ep.* vol. XII (1958) Indices pp. 33 f. and 145 f. P. S. Allen, 'Erasmus' Services to Learning', *Proceedings of the British Academy* (1924/5) (Annual Lecture on a Master Mind) and *Erasmus* (1934).—The position of Erasmus's editions in the history of biblical scholarship has often been treated; see *The Cambridge History of the Bible*, vol. II: 'The West from the Fathers to the Reformation', ed. G. W. H. Lampe (1969) pp. 492 ff. Cf. also Bruce M. Metzger, *The Text of the New Testament* (2nd ed., 1968) pp. 98 ff.

'examined' new manuscripts, that is pupils read them aloud to him in the afternoon and he made notes. We see how haphazard the procedure was. For Revelation he had obtained only one Greek manuscript from Reuchlin; unfortunately a few verses at the end were missing, and Erasmus quietly translated them from the Latin into Greek himself. This was the text that, printed in Robertus Stephanus's folio Greek Testament of 1550, became the *textus receptus*; three centuries were to elapse before it was discovered that there was no authority for the Greek wording except Erasmus's knowledge of the Greek language.[1] This curiosity may be mentioned by the way, but it remains true that Erasmus's Greek New Testament is his greatest humanistic work. Not only did it win the sanction of the Medicean pope Leo X, who acknowledged it as an outstanding scholarly service to the Roman church. Immediately after its publication it was also used by Luther in his revolutionary lecture of 1516 on the Epistle to the Romans,[2] and the Erasmian text became the main, though not the only source for Luther's German translation of the New Testament.[3] Soon afterwards Erasmus was exposed to heavy attacks from both sides. His intellectual superiority, however, enabled him to remain a faithful member of the Catholic church without giving up his vigorous criticism and his individual views. We may compare the attitude of Lord Acton, who after the first Vatican Council in 1870 did not follow his Munich teacher I. Döllinger in leaving the church. And it looked as if something of the spirit of Erasmus was going to be revived in the second Vatican Council, at least in its beginnings. Pope John XXIII,[4] praising the 'materna vox' with deep love and understanding, included among the Latin books he recommended for the education of young priests those of Erasmus, Vives, and Pontanus.

New editions of the classics were not so urgently needed, thanks to the work of Italian humanists. Nevertheless, Erasmus produced editions[5] of Terence, Curtius, Suetonius, Pliny's *Natural History*, Livy, and of a few philosophical writings of Cicero, he translated[6] parts of Xenophon, Plutarch, Galen, Lucian, and Euripides' *Hecuba* and *Iphigenia*, and he published Greek texts of Aristotle and Demosthenes. No small

---

[1] Lagarde, *GGA* (1885) p. 64.

[2] *Werke* (Weimarer Ausgabe) vol. LVI, p. lii (Register).

[3] W. von Loewenich, 'Die Eigenart von Luthers Auslegung des Johannes-Prologes', *SB Bayer. Akad., Phil.-histor. Kl.* 1960, H.8, p. 27.

[4] See *Constitutio* 'Veterum Sapientia' (1962) and cf. *Gymnasium* 71 (1964) 200 and 203.

[5] *Ep.* vol. XII (1958) Indices pp. 30 ff.

[6] On translations see Waszink, 'Einige Betrachtungen über die Euripidesübersetzungen des Erasmus und ihre historische Situation', *Antike und Abendland* 17 (1971) 70 ff.

achievement for a parergon. But the essential need, beside the Greek Testament, was to open the way to the original text of the Church Fathers. Erasmus had been interested in St. Jerome from his time at school[1] and in the monastery, and he began working at the letters in 1500. The text of the complete work went through the press at Basle side by side with the New Testament and was published in the same year, 1516, in nine large folio volumes; it was reprinted twice by Froben and once in Paris. He also published texts of Cyprian, Hilary, Ambrose, Irenaeus, Augustine (completed in 1529 in ten stout folio volumes), and John Chrysostom. But he undoubtedly regarded Jerome, and later Origen, as being of unique importance. He had begun to work on the text of Origen, as he had on that of Jerome, shortly after 1500, and to the completion of the task he devoted his waning strength at the end of his life; two folio volumes appeared three months after his death in 1536. The reason for the particular respect that Erasmus felt for Origen is to be found in the similarity of the problems that faced the two men. Erasmus in the sixteenth century was trying to construct a synthesis of Christianity with culture and to create a new biblical philology, as Origen had done in the third century A.D.

If we brought together all these editions of Erasmus, it would be a mountain of volumes. We can hardly imagine how difficult it was to explore the world of manuscripts at that time and to make careful collations. Later editors usually complain that Erasmus did not make sufficient use of manuscript readings, but relied too much on conjectures. Few modern scholars have taken the trouble to consider Erasmus's actual intentions and to examine his editing in detail; but one of the greatest experts, J. de Ghellinck,[2] finally achieved a favourable and just assessment of the especially large and difficult edition of Augustine. Erasmus did not aim at an absolute precision of text for specialists or seek to make his annotations and paraphrases an exercise of pure learning. The introductory essays to the New Testament have already been mentioned, the so-called 'Methodus'. But in a certain sense, the numerous books which accompanied his editions were also instructions to the reader for the understanding of the language and the subject-matter of the Greek and Latin texts. It should perhaps be noted that in making this distinction between running commentaries and monographs Erasmus was in the middle of a long and continuous

[1] See above, p. 69.

[2] On Erasmus's edition of Augustine and on other important editions of Church Fathers with a just appreciation of the work of editors and printers see J. de Ghellinck in *Miscellanea J. Gessler* 1 (1948) 530 ff.; cf. also below, p. 83 n. 3 on the Amerbachs.

tradition. Aristarchus' ὑπομνήματα and συγγράμματα were the earliest examples in Hellenistic times;[1] but there are others still to be observed in the twentieth century, for instance in the writings of Wilamowitz.[2]

A few of Erasmus's monographs may be listed: *De ratione studii* and *De duplici copia rerum et verborum*, both in 1511, dedicated to John Colet for St. Paul's School, the *Ciceronianus* (1528) on Latin style,[3] and in the same year *De recta Latini Graecique sermonis pronuntiatione*;[4] in both these writings on style and pronunciation the discussions of several generations reached their culmination. Even the *Colloquia*, published 1516, were intended first 'ad linguam expoliendam', and secondly 'ad vitam instituendam'. But this collection of extremely witty satires in dialogue form on types of the social life of the age, with some bitter attacks on certain doubtful figures of the clergy, was ultimately read by the general public of the whole world and made Erasmus's reputation as a brilliant, even thrilling writer. His boldest satire, the *Praise of Folly*, *Encomion Moriae*,[5] was written in the house of Thomas More in London (1509) and dedicated to him. It is a model of perfect Latin and was intended to be exemplary in form and contents: lack of knowledge, ignorance, stupidity are shown to be the cause of all evils in this world and of all sins against God's laws. Scholarship, of course, helps to build up a bulwark against ignorance. So he set out to teach the various forms of knowledge in a number of monographs, the *Enchiridion militis Christiani* (1501), the *Institutio principis Christiani* for the young Charles V (1516), the *Ecclesiastes* (1535), which he published in his last year in Basle; it was to have been dedicated to Cardinal John Fisher for his theological school at Cambridge, but just a few weeks before Erasmus had finished the book John Fisher and Thomas More died as martyrs. Of his many friends in many countries Thomas More was in Erasmus's eyes the perfect model of humanity in this world, 'optimum exemplar' or even 'exemplar absolutum'; the highest spiritual culture, combined with moral strength and an active devotion to the community, the *civitas*, made him in the words of a classical phrase that Erasmus applied to him 'omnium horarum homo'.[6] It is a further notable coincidence that More's *Utopia*

---

[1] See *History* [1] 213.   [2] Cf. *Ausgewählte Schriften* p. 272.

[3] Cf. above, pp. 43 f. and 53 f.

[4] On pronunciation of Greek cf. above, p. 53, and on Reuchlin below, p. 88.

[5] Erasmus, *Praise of Folly*. Translated by Betty Radice with valuable notes by A. H. T. Levi (1971).

[6] *Ep.* 1233.94 (Sept. 1521 to Budé); Sueton. *Tib.* 42.3 and Quintil. VI 3.111; Allen gives no references to the ancient sources in this case. 'A man of all seasons' became a commonplace in English (see R. W. Chambers, *Thomas More* (1935) p. 177) and even the title of a play by Robert Bolt in 1960.

and Erasmus's first edition of the New Testament as well as his *Colloquia* were all published in the same year.

Erasmus also wrote a series of polemic treatises, of which two are particularly relevant to our subject. The *Antibarbari*[1] is a dialogue in which Erasmus and some friends of his defend the humanistic programme against its opponents, called 'Barbarians'. The first edition was printed in 1520, but a manuscript in Gouda has preserved an earlier draft, probably of 1495. A comparison of the two clearly demonstrates that in spite of all verbal differences there is a definite unity in the thought and the work of Erasmus. The *De libero arbitrio* διατριβή (1524)[2] is directed against Luther, who seemed to be the most dangerous opponent of Erasmian humanism. The principle of Erasmus was the awareness and recognition of free will and of full responsibility for all one's actions;[3] without that all his theories and the whole work of his life would be senseless. Nothing was more strongly opposed to this principle than Luther's doctrine of *necessitas absoluta*. The consequence of Lutheranism he believed to be 'tumultus, dissidium, seditio, factio'; Luther's nature and the spirit of his followers seemed to him 'saevus, austerus, ferox' and even 'atrox'. (The sinister history of the term 'atrocity' seems to start from these polemics.) Erasmus's desire was for peace and harmony, reconciliation of knowledge and belief, liberty and piety, freedom and the Church. The Erasmian idea of 'tranquillitas' is connected with his 'humanitas'. He very often wrote about peace. The most famous part of his widely read *Adagia*, a collection of the wisdom of the ancients[4] dedicated 'philologis omnibus', is 'dulce bellum inexpertis',[5] often printed as a separate pamphlet, the best ever written against aggression in the same spirit as Thomas More's *Utopia*. Nobody hated war more than Erasmus, but oddly enough, in one of the most dangerous years of Turkish invasion (1529) he was forced to write *De bello Turcis inferendo*, arguing that it would be a crime not to defend

---

[1] See *Ausgewählte Schriften* (above, p. 71 n. 4). The text of the Gouda manuscript was discovered by P. S. Allen, *Erasm. Ep.* vol. v (1924) p. xx and published by Hyma, *The Youth of Erasmus* (1930, 1968²), see above, p. 69 n. 1. In my article on the *Antibarbari* I politely drew attention to inaccuracies in Hyma's *editio princeps*; he called this a 'savage attack on the method' (p. 384 of the second edition). It was disappointing to see that he did not bother to correct the elementary mistakes in reading and printing the text of the Gouda manuscript and of the first printed edition; but it does not matter any longer, as the second editor of the *Antibarbari* in *Opera omnia* I, 1969 (see above, p. 71 n. 1) established the text most carefully, using also my readings and suggestions.

[2] Hrsg. von Johannes von Walter, *Quellenschriften zur Geschichte des Protestantismus* 8 (1910, reprinted 1935).

[3] See above, p. 40 (Valla).   [4] See *History* [I] 83 f., 208 f.

[5] It is an ancient saying of Pindar's, fr. 110 Sn. γλυκὺ δὲ πόλεμος ἀπείροισιν.

Western Christian civilization against barbarous aggression from the East and that the Christian peoples in Europe must be reunited in a new crusade, in which the true 'miles Christianus'[1] will be victorious.

Erasmus used his scholarship and humanism for the promotion of the universal Church, that it might retain the spiritual leadership of the whole of Christendom which the Church of his day was in danger of losing. At the end of his life he saw it endangered from the inside and from the outside. His desire had been to bring unity out of all conflicts: princes and priests and intelligent laymen should be educated to unite all men in the fellowship of Christ. But instead of unity there came division; he clearly foresaw what a tragedy final disruption would be. He could not do more than use his spiritual armour as a 'miles Christianus'. He did not fail, as is so often said, especially by historians; the others failed to accept his warning. And what we may call the tragedy in Erasmus's life lies in the fact that he was to a large extent misunderstood by the Church,[2] to which he had devoted all the vigour of his versatile mind.

[1] See above, p. 77.
[2] K. Schätti, 'Erasmus von Rotterdam und die Römische Kurie', *Basler Beiträge zur Geschichtswissenschaft* 48 (1954), has collected the material.—Ignatius of Loyola and the rising order of the Jesuits could not feel sympathy with Erasmus's 'Philosophia Christi': cf. R. Pfeiffer *Humanitas Erasmiana* (1931) p. 22, A. Flitner, *Erasmus im Urteil seiner Nachwelt* (1952) p. 86, H. Rahner, *Ignatius von Loyola als Mensch und Theologe* (1964) pp. 373 and 512 f.

VIII

AUTOUR D'ÉRASME[1]

It has often been said that Erasmus's work on the whole ended in failure. We have argued[2] that it was not he who failed, but others who failed to accept his timely warnings. In any case, his influence was immensely strong, both immediately on his own and on the next generation, on his many friends and his few pupils (since he only occasionally acted as professional teacher), and also through his writings on the European mind and on scholarship for centuries. One of the chief criticisms made against humanism is that the revival of the Greek and Latin classics checked the natural development of the European nations, especially in the north: a gap, it is said, was opened between an upper class, now educated in foreign languages and ideas, and the common people. But where in the fifteenth century, may we ask, was the unity which could be thus divided? Its existence is obviously a romantic illusion. When we turn from national unity to European universalism, what destroyed it was the growing national selfishness and the religious revolutions of the sixteenth century, and only the humanists of that time formed a supranational group of kindred spirits which tried to save Europe from complete cultural disruption. Erasmus's work was most effective in creating and strengthening connecting links. And it was on the Erasmian model, as we shall see, that true scholarship prospered,[3] not on that of the biblical exegesis of the reformers and the new Protestant Scholasticism, still less in the narrow traditionalism of the Catholic counter-reformation.

On the Upper Rhine, in Alsatia, Switzerland and Baden, Erasmus had a large circle of followers.[4] His most devoted and faithful pupil was

---

[1] I borrow this title from L. Bouyer, *Autour d'Érasme, Études sur le christianisme des humanistes catholiques* (1955).

[2] See above, p. 81.

[3] *Philologia perennis* (1961) p. 13 'Diese Einheit der Philologie' and the following sentences are open to misunderstanding in so far as they seem to imply that the Erasmian line is identical with the Roman Catholic doctrine.

[4] G. Ritter, 'Erasmus und der Humanistenkreis am Oberrhein', *Freiburger Universitätsreden* 23 (1937); E. W. Kohls, 'Die theologische Lebensaufgabe des Erasmus und die oberrheinischen Reformatoren', *Arbeiten zur Theologie* 1. Reihe, Heft 39 (1969).

Beatus Rhenanus,[1] an Alsatian from Schlettstadt (1485–1547), who after studying in Paris resided at Basle from 1511 to 1527, for many years simultaneously with Erasmus,[2] working there for the great printers and publishers, especially for Johannes Froben and the Amerbachs.[3] After Froben's death he returned to his native city, to which he left his very remarkable library still preserved almost undamaged.[4] Other humanists from Petrarch onwards had tried to bequeath their books for public use, but Beatus Rhenanus was the first to succeed. It was through him that Froben asked Erasmus for the *New Testament*: 'Petit Frobenius Novum abs te Testamentum habere.'[5] Rhenanus wrote the first biography of Erasmus and probably made the first complete edition of his works (9 volumes, Basle 1540), on which the later collection (Leyden 1703–6) was based. Basle had been the principal home of humanism in the north since the Council,[6] and its printers continued to issue classical texts as well as collected works of the great humanists, Petrarch and his followers;[7] many of these editions are still indispensable to us, as we have seen.

The cities on the Upper Rhine became not only a centre of printing presses and new libraries, but also an important meeting-place for educationalists and schoolmasters. Jacob Wimpfeling (1450–1528)[8] was educated in the school of Schlettstadt and settled down finally in Strasburg, where he tried to reform the system of education and to introduce better modern school-books on grammar and style. His achievement was outdone by the younger Jacob Sturm (1507–89),[9] headmaster of the Gymnasium in Strasburg for forty-three years. He had gone through the schools of Liège and Louvain which were under the influence of the 'Devotio Moderna' and of Erasmus, and promoted

---

[1] See Allen's introduction to Er. *Ep.* 327, and *Ep.* XII pp. 49 f. Indices s.v. Beatus Rhenanus. *Briefwechsel* des Beatus Rhenanus hrsg. v. A. Horawitz und K. Hartfelder (1886, reprinted 1962).

[2] See above, p. 73.

[3] *Die Amerbach-Korrespondenz*, hrsg. v. A. Hartmann, 6 vols. (1942–67), a wonderful edition with indexes to each volume. See also H. Thieme, 'Die beiden Amerbach. Ein Basler Juristennachlass der Rezeptionszeit', *L'Europa e il diritto Romano. Studi in memoria di P. Koschaker* I (1954) 137 ff.

[4] H. Kramm, *Deutsche Bibliotheken* (below, p. 87 n. 3) pp. 102 f.

[5] Er. *Ep.* 328.36 (17 Apr. 1515), a letter characteristic of the learned writer, as it is full of variant readings of Seneca's *Lucubrationes*, at that time in Froben's press.

[6] See above, p. 60.

[7] Friedrich Luchsinger, 'Der Basler Buchdruck als Vermittler italienischen Geistes', *Basler Beiträge zur Geschichtswissenschaft* 45 (1953) 115 ff. and P. Bietenholz, 'Der italienische Humanismus und die Blütezeit des Buchdrucks in Basel 1530–1600', ibid. 73 (1959) 10 ff.

[8] P. S. Allen on *Ep.* 224.

[9] W. Sohm, *Die Schule Johannes Sturms und die Kirche Strassburgs . . . 1530–1581*, Hist. Bibl. 27 (1912).

in his writings and in the practice of his school the ideal of 'sapiens et eloquens pietas', stressing the word 'eloquens'. A new oratory modelled on Cicero was to be cultivated and to be put at the service of the new Protestant piety. The preacher must become the perfect orator, preaching the genuine doctrine of Christ in the purest and most beautiful language.

Rhenanus had no ambition to appeal to a wider public; he was a quiet, steady, industrious worker, confining himself to pure learning. When we compare his edition of Tacitus' *Germania* (1519) and his *Rerum Germanicarum libri tres* (1519) with Conrad Celtis's works on the same subject (1515),[1] we are struck not only by the contrast between the two humanists, but also by the enormous and immediate effect of Erasmus's scholarship on those around him.

Rhenanus's general conception of history was that early Germany had been without culture until civilization was introduced, together with Christianity, as a legacy of the peoples of the ancient world. On the other hand, the decline of civilization in Italy in the fifth and sixth centuries was due not to 'Gothic' invasions, as Valla[2] had supposed, but to the growth of indifference and stupidity. Historical and literary terms, familiar from the later sixteenth century onwards, seem to have occurred first in Rhenanus's writings: 'media antiquitas',[3] for instance, for the time between the end of antiquity and the revival of learning, and 'classici' for the writers of the first class.[4]

Beatus Rhenanus possessed a fuller knowledge of the writers on later Roman history and the German peoples than earlier scholars had, and he tried to give a careful interpretation of their texts. He was surprisingly successful in his attempts to restore corrupt passages in Tacitus and Ammianus Marcellinus.[5] 'Labor et animus', as he said, are necessary, but useless without 'iudicium'.

---

[1] See above, p. 64.   [2] See above, p. 35.

[3] P. Lehmann, 'Vom Mittelalter und von der lateinischen Philologie des Mittelalters', *Quellen und Untersuchungen zur lateinischen Philologie des Mittelalters* v 1 (1914) 3, 6 f.

[4] Cf. *History* [1] 207. Cf. G. Luck, 'Scriptor classicus', *Comparative Literature* 10 (1958) 150 ff. The earliest modern example of the word 'classicus' known to me occurs in 1512 in the *Briefwechsel* des Beatus Rhenanus (above, p. 83 n. 1) no. 25: 'classici auctores'. B. Kübler, *RE* III 2629.20 ff. refers to Melanchthon's dedication of Plutarch's Λάθε βιώσας (1519) 'Plutarchi . . . classici videlicet auctoris'. In any case, the ancient term was revived in the circle of Erasmus; I have not found it in his own writings. One of his Spanish correspondents correctly included St. Augustine among the 'classici': *Ep.* 2003.33 (29 June 1528) 'solus . . . is auctor ex classicis reliquus videbatur qui nobis Erasmo obstetricante renasceretur' (Fonseca, Archbishop of Toledo). See also above, p. 12.

[5] One of his generally accepted conjectures in Ammianus Marcellinus XVIII 2.15 was the starting-point for E. Norden, *Alt-Germanien* (1934) pp. 11 ff., who rejected Rhenanus's reading, but gave an appreciation of his merits as editor, critic, commentator, and historian.

By Beatus Rhenanus's time almost every Latin text known to us had been discovered and published in print.[1] But he still found a new one in 1515 in the Alsatian monastery of Murbach, Velleius Paterculus, of whom he made the first edition in 1520; it is *codicis instar*, as the manuscript was subsequently lost. The last discovery in this field, and a more important one, was made in 1527, when Simon Grynaeus came upon five books of the fifth decade of Livy (41–5) in the monastery of Lorsch. Grynaeus (who was also a member of the circle of Erasmus) was able to make use of the find when he edited Livy in 1531. But the definitive edition was made in 1535 by Beatus Rhenanus and Sigismund Gelenius, who used not only the new text of the fifth decade from Lorsch, but also two new manuscripts of the fourth decade from Worms and Speyer (which were afterwards lost). Gelenius, a member of this very active group of scholars, also had a new manuscript from Hersfeld of the later part of Ammianus Marcellinus[2] (now lost like the other manuscripts just mentioned, except that six folios were rediscovered by chance in 1876), and his edition of 1533 has ample contributions by his friend Rhenanus, whose output as a whole was very great indeed.

At least two further members of the large Erasmian circle on the Upper Rhine should be mentioned, the Swiss humanist and poet Heinrich Loriti (1488–1563) from Glarus, called Glareanus, and Ulrich Zasius (1461–1535) born in Constance,[3] who revived the study of Roman law in Germany. Glareanus was one of those classical scholars who established their reputation by composing Latin poèms: he received the poet's laurel in 1512 from the emperor Maximilian.[4] As a poet himself, he naturally edited and commented on many of the Roman poets. But his special interest was in problems of chronology, mathematical geography, and even ancient musical theories; in fact, the *Dodecachordon*, published in 1547, may be regarded as his most

[1] Cod. Medic. I of Tacitus was found in 1508 in Corvey, Plin. *Ep. ad Trai.* in 1500 in Paris.

[2] Cf. Pasquali, *Storia* pp. 81 ff.

[3] See Er. *Ep.* XII Indexes pp. 102 f. on Glareanus and p. 188 on Zasius.

[4] This had the understandable and amusing consequence that he never published an epic poem on the victorious battle of his compatriots against the Austrians at Näfels which he had just written in 1510; it seems to be preserved only in a handwritten copy made by Glareanus's pupil J. E. von Knöringen which I found by chance attached to a printed edition of Glareanus's elegiacs of 1516 (now Clm. 28325). See 'Neues von Glareanus', *Zentralblatt für Bibliothekswesen* 34 (1917) 284 ff. with further references. When I had written this note, I had an opportunity of asking Meinrad Scheller about that Latin poem and he immediately gave me an offprint from the *Jahrbuch des Historischen Vereins des Kantons Glarus* 53 (1949) 1–36 'Henrici Glareani carmen de pugna confoederatorum Helvetiae commissa in Naefels', hg. von Konrad Müller, a critical edition of the text of Knöringen's copy (in the original publication on pp. 58–119; cf. also p. 9 n. 9 with reference to the *Zentralblatt* of 1917).

important work.[1] Erasmus seems to have been impressed by the wide range of his learning and his industrious teaching activity in Basle and in Freiburg, and even went so far as to apply to him the formula 'omnium horarum homo'.[2] But a certain pomposity in his style and behaviour was not to Erasmus's taste.

Zasius,[3] on the other hand, became an intimate, beloved friend of Erasmus, who was always a welcome guest in the learned lawyer's Freiburg home. In one of his most charming letters the elderly Erasmus, with his constant gastric troubles, tells how he was served a chicken for his dinner on a Friday night and how they were both denounced: 'corycaeus sycophanta nidorem eius pulli detulisset ad Senatum';[4] it was typical of him to treat a delicate question, that of abstinence on Friday, like this in an ironical, but fundamentally serious style.

Zasius was called an 'alter Politianus'[5] by Erasmus, not without some exaggeration. His discoveries of Roman legal manuscripts in the monastery of Murbach were the most important before Niebuhr's, and he was a true interpreter of the ancient texts; he practised also as a lawyer in Freiburg, putting his knowledge of Roman civil law to good use. Zasius's publications show how effective Erasmian ideas could be in the province of jurisprudence.[6] Erasmus's general thinking about justice and law was, of course, dependent on his concept of 'Christi philosophia',[7] and it made a considerable impression on his contemporaries. The effect even extended to Italy, where Andrea Alciato (1492–1550),[8] who introduced the study of jurisprudence into France, was regarded as an Erasmian.

Although the group of scholars on the Upper Rhine, like Erasmus himself, were chiefly Latinists, some of them knew Greek fairly well. The first great exponent of Greek studies in Germany,[9] and the first to

---

[1] Reprinted in *Publikationen älterer praktischer und theoretischer Musikwerke*, herausgegeben von der Gesellschaft für Musikforschung, vol. 16 (1888).

[2] See above, p. 79, on Thomas More.

[3] *Opera*, 6 vols. (Lugd. 1550), *Epistulae* (1774), more in Er. *Ep.* and *Amerbach-Korrespondenz* (above, p. 83 n. 3); monograph by R. Stintzing, *U. Zasius* (1857); G. Kisch, 'Erasmus und die Jurisprudenz seiner Zeit', *Basler Studien zur Rechtswissenschaft* 56 (1960) 317 ff. with bibliography p. 318.2.

[4] Er. *Ep.* 1353.7 (23 Mar. 1523); cf. his essay on 'Ichthyophagia' (1526), *Opp.* I 805 B–E.

[5] On Politian's legal studies see above, p. 43.

[6] We shall have to say a little more about this subject in the following chapter on France, see P. Koschaker, *Europa und das Römische Recht* (2nd ed. 1953) *passim*; but this is a very wide field of special research on which I do not feel competent to give more than occasional hints.

[7] See G. Kisch, 'Erasmus und die Jurisprudenz seiner Zeit' (above, n. 3) *passim*.

[8] Ibid. pp. 304 ff.

[9] There is no justification for calling the Dominican John Cono (or Kuno) of Nuremberg (1463–1513) the 'true founder of Greek studies in Germany', as Geanakoplos (above, p. 56

link Oriental (that is, Hebrew) studies with them, was Johannes Reuchlin, a native of Pforzheim in Baden (1455–1522);[1] it is characteristic that instead of Latinizing his name, he turned it into Greek Καπνίων, from καπνός, 'Rauch'. Erasmus celebrated his memory after his early death in *De Capnionis apotheosi*. Reuchlin learnt his Greek, like Erasmus and others, from Greek immigrants in Paris; and when he came to Rome in order to continue his studies, Johannes Argyropoulos,[2] one of the Greek scholars in Italy, exclaimed: 'Through our exile Greece has flown across the Alps.' In Italy Reuchlin also came into contact with the Florentine Neoplatonists and mystics who studied the cabbalistic tradition, and on a later visit to Rome he began to learn Hebrew; by 1506 he was able to publish a book, *De rudimentis linguae hebraicae*, with a grammar and dictionary, and could justly claim to be the first of the Latins to do such a thing. He was in diplomatic service, was a judge in his own country, a peaceful gardener, a successful writer of Latin comedies, and a passionate collector of books. He bequeathed his library,[3] which contained nearly all the Greek texts printed in Italy, to his native city of Pforzheim, just as Rhenanus had left his to Schlettstadt. In his youth, and again in the last years of his life, he acted as a lecturer on Greek and Hebrew in the universities of Ingolstadt and Tübingen; otherwise he held no university post. The German universities were still reluctant to admit the new learning, and none of the humanists so far mentioned had yet been offered more than a temporary position by some universities.

Reuchlin's part as an intermediary in the field of Greek studies has no parallel in other countries. It is a surprising fact that exiled Greek scholars did not come to Germany to teach their language, as they did

---

n. 3) p. 136, after others is inclined to do. Having attended lectures of Musurus in Padua, Cono was able to teach Greek in Basle from 1505 onwards, to become tutor in the house of the Amerbachs and corrector in their printing office. Erasmus and other contemporaries expressed their approbation of his merits, but no one would have compared them with those of Reuchlin. See references in Erasm. *Ep.* xii Indices (1958) 120 and *Amerbach-Korrespondenz* i Index (1942) 478; cf. M. Sicherl, 'Nürnberg und der griechische Humanismus in Deutschland', *Jahres- und Tagungsberichte der Görres-Gesellschaft* (1971) pp. 39 ff.

[1] 'Joh. Reuchlins Briefwechsel', hrsg. von L. Geiger, *Bibliothek des litterarischen Vereins in Stuttgart* 126 (Tübingen 1875); cf. Erasm. *Ep.* i 555. *Zeitschr. f.d. Geschichte des Oberrheins* 76 (1922) 249–330. *Reuchlin. Festgabe seiner Vaterstadt Pforzheim zur 500. Wiederkehr seines Geburtstages* hrsg. v. M. Krebs (1955). For a critical review of modern literature on Reuchlin see H. Goldbrunner, 'Reuchliniana', *Archiv für Kulturgeschichte* 48 (1966) 403 ff.

[2] 'Ecce Graecia nostro exilio transvolavit Alpes', quoted by Melanchthon, 'Declamationes', *Corpus Reformatorum* xi 238 and 1005.

[3] K. Christ, 'Die Bibliothek Reuchlins in Pforzheim', *Beihefte zum Zentralblatt für Bibliothekswesen* 52 (1924). H. Kramm, 'Deutsche Bibliotheken unter dem Einfluss von Humanismus und Reformation', ibid. 70 (1938) 266 f. K. Preisendanz, 'Die Bibliothek J. Reuchlins', *Festgabe* (above, n. 1) pp. 35 ff.

at first in Italy and then on a larger scale, as we shall see, in France. The rich free cities of south Germany, such as Augsburg or Nuremberg had many links with the Italian seats of humanism, especially Venice, where the great Cretans, Musurus and others, were living and teaching. But none of these scholars was invited across the Alps; only commercial travellers came in the course of time to sell their manuscripts. The gradual change in this situation seems to me to have been entirely due to Reuchlin. One of his pupils, Georg Simler, the author of an elementary Greek grammar for German students, was also the first to instil the knowledge of Greek into Reuchlin's great-nephew Philipp Schwarzerd, better known as Melanchthon (1497–1560); like his great-uncle, who took the keenest interest in his gifted young relative, Schwarzerd chose to turn his name into Greek. His gifts were not exactly those of a scholar, but as an educationalist he was a genius. At the age of twenty he produced his *Institutiones linguae Graecae*, printed in 1518, the Greek grammar which made Reuchlin's ideas, especially those on the pronunciation of Greek, popular and was used in German schools for three centuries.

The controversy about pronunciation has always been associated with the name of Reuchlin. As the first German teacher of Greek he introduced into Germany the pronunciation of the modern Greeks, his teachers in Italy, the so-called Reuchlinian or itacistic pronunciation in which the vowels η ι υ and the diphthongs ει οι υι were all pronounced like the Italian *i*. Erasmus,[1] who was a great friend of Reuchlin and an admirer of his scholarship, protested against this in his treatise of 1528 *De recta Latini et Graeci sermonis pronuntiatione*. Comparing the deformed pronunciation of Latin in modern national languages, he explained that the simplified system of vowels in modern Greek could not be the original one. He showed how Latin had adapted the Greek vowels as well as consonants in ancient times, and thus demonstrated the original difference of vowels like η ι υ and so on. Characteristically Erasmus did not do this in a dry article; on the contrary, he chose the form of a witty dialogue between a lion and a bear in which it is shown how the lion's whelp should learn to read Greek properly with amusing examples of how badly Greek was read by Dutch, Scotch, German, or French people. The history of the problem is not as simple as the traditional distinction between Reuchlinians and Erasmians makes it appear. Erasmus had predecessors, and he was no doubt fully aware that others before him had seen that classical Greek needed a different pronunciation

[1] Cf. above, pp. 53 and 79.

from that of modern Greek. He recommended inviting Janus Lascaris to the Collegium Trilingue of Louvain because of his correct pronunciation of the Greek language.[1] The so-called Erasmian pronunciation was generally adopted in Western Europe,[2] but in the Protestant as well as in the Catholic part of Germany and in Italy the Reuchlinian practice prevailed until the time of German New Humanism. In respect of the stress accent there was no difference between the two parties: both stressed the syllables on which the accent fell.[3]

A strange event in the life of Reuchlin became a test case for the division into humanists and anti-humanists in Europe. The starting-point for the conflict in which he was involved, however, was not in his classical, but in his Hebrew studies. The university of Cologne, where the Dominicans held the principal posts, was a stronghold of the conservatives; it supported Johannes Pfefferkorn, who had written four diatribes against Jewish books between 1507 and 1509, and decided that all books in Hebrew should be confiscated and destroyed. Reuchlin protested against the wholesale destruction of this literature, arguing that only books expressly anti-Christian and therefore dangerous should be forbidden. A battle of the books ensued. In 1511 the Dominicans applied to the Imperial court; but Reuchlin, the only Oriental scholar among the lawyers, handed in a written opinion in which he exposed the ignorance and fanaticism of his opponents. Accused of heresy in 1513, he appealed to the pope, fortunately at that time the Medicean Leo X. Reuchlin was able to hand in a memorial to the papal court in which his orthodoxy was certified by the Emperor himself, by the king of France, princes, electors, bishops, abbots, and Swabian towns; all the cardinals were inevitably involved in the controversy on the other side. The excitement was universal, especially throughout Germany. But there was no decision yet from Rome. In 1514 Reuchlin's friends, famous scholars from all parts of Europe, published a collection called *Illustrium virorum epistulae ad Ioannem Reuchlinum*; it was followed in the next year by a volume bearing the title *Epistulae obscurorum virorum ad . . . Ortwinium Gratium*. This Ortwin Gratius (1491–1551), depicted here as representative of the obscurantists, was a real member of the Faculty of Arts at Cologne; but his correspondents were fictitious young

---

[1] Er. *Ep.* 836.10 'germanam Graeci sermonis pronuntiationem'.

[2] E. Drerup, 'Die Schulaussprache des Griechischen von der Renaissance bis zur Gegenwa:t' I (1930) II (1932), *Studien z. Geschichte u. Kultur des Altertums, Ergänzungsbd.* 6.7.

[3] The curious practice of applying the Latin rule to the Greek words (anthrópos, árete) was introduced by a Dutch scholar Henninius in the later seventeenth century; schools in Holland and in its neighbouring countries seem to have used it.

graduates of various German universities. The *Epistulae* are a clownish parody of the casuistry and ignorance of these people; the language is immensely amusing, a grotesquely Latinized German idiom, carried through with ingenious consistency, totally different from the 'Latinum culinarium' ridiculed in Valla's pamphlets against Poggio.[1] In 1516 an appendix to the first volume and a new second part appeared. A convincing philological analysis[2] has demonstrated that the first part, a satirical masterpiece, was written by Crotus Rubianus, a member of the distinguished group of humanists in Thüringen, of which Mutianus Rufus (1471–1526), a school-fellow of Erasmus at Deventer, was the acknowledged head; the second part of the *Epistulae*, more - in the aggressive style of the ancient ἴαμβος, was the work of Ulrich von Hutten. When the papal court finally made a decision, not entirely favourable to Reuchlin, he was no longer under attack.

Nothing more clearly illustrates the unity of the humanistic movement in and outside Germany, despite all personal, local, national, and religious differences, than the short history of the *Epistulae obscurorum virorum*. It is a sort of joyful and sportive comedy, before in Erasmus's words 'tragoedia incipit'.

In precisely the same years, 1515/16, Luther started his lectures in Wittenberg on St. Paul's Epistle to the Romans,[3] and in the next year 1517 the open struggle began. Interpreting Rom. 1 : 17 from his own deep religious experience, Luther believed that he had found there the expression of justification by faith alone, an interpretation which, despite its consequences, he believed to be in harmony with that of St. Augustine. This was Luther's fundamental conception, and through further exegesis he came to hold it with increasing conviction. Former interpretations, the traditions and doctrines of centuries, became superfluous, since the scriptures were directly understandable: 'scriptura sacra sui ipsius interpres'[4] had a new sense. So thought the other reformers also. Luther reproached Erasmus: 'Novum Testamentum

---

[1] See above, pp. 35 f.

[2] The analytical research was initiated by W. Brecht, 'Die Verfasser der Epistolae obscurorum virorum', *Quellen und Forschungen zur Sprach- u. Culturgeschichte der germanischen Völker* 93 (1904). Brecht used the text in the Supplementum to the *Opera* of U. von Hutten, ed. E. Boecking (1870); a new critical text was published by A. Bömer (1924) who partly differed from Brecht and dealt with the later discussions of the analytical question. See also H. Holborn, *U. von Hutten* (1929).

[3] Cf. above, p. 77; see W. Grundmann, *Der Römerbrief des Apostels Paulus und seine Auslegung durch Martin Luther* (1964) with references to the critical editions.

[4] Cf. above, p. 76.

transtulit et non sensit'—missing the personal experience, he had failed to achieve a truly religious understanding.[1]

In respect of the 'pagan' Greek and Roman literature Luther never ceased to enlarge his knowledge.[2] It was characteristic of him to put the fables of Aesop beside the Bible, saying that 'there is no better book in secular pagan wisdom' ('kein feineres Buch in weltlich heidnischer Weisheit'). Luther even conjectured in the preface of his German translation of a number of *Aesopea* that 'perhaps no man on earth had ever been called Esopus' ('vielleicht nie kein Mensch auff Erden Esopus geheißen . . . die Fabeln seien Jar zu Jar gewachsen und gemehrt . . . und schließlich gesammelt worden').[3] Apparently none of the humanists who were interested in the ancient fables, Poggio, Valla, Erasmus, had expressed this opinion which became quite common in the nineteenth century.

Erasmus, Reuchlin, and Luther were straightforward in their thinking; compared with these, Reuchlin's great-nephew Melanchthon[4] was inclined to complicated and questionable compromises. He tried to bridge the chasm between the humanists and the religious reformers, honestly, but in vain. He was born in 1497 in Bretten, not very far from Reuchlin's birth-place, Pforzheim. An infant prodigy, he became a Master of Arts at the age of sixteen and began to lecture in the university of Tübingen at twenty. A fervent admirer of Erasmus, who admired him equally in return, he grew up in Erasmian ideas. After the publication of his Greek grammar[5] he was recommended by Reuchlin to the Elector of Saxony for a chair of Greek in the university of Wittenberg in 1518. His first lecture was on Homer, the second on the Greek text of St. Paul's Epistle to Titus, the third on the Hebrew text of the Psalms—a combination, one might think, of the interests of Erasmus and Reuchlin. Now, however, he met Luther in Wittenberg. The impression of his personality and of his religious conviction was over-whelming, and Melanchthon's flexible mind was immediately captured. But he did not abjure humanism; he tried to compromise. The way of

---

[1] On Luther and Lutheranism from Erasmus's point of view see also above, p. 80. See also K. Holl, 'Luthers Bedeutung für den Fortschritt der Auslegungskunst', *Gesammelte Aufsätze* I ([7]1958) 544–82; W. Bodenstein, 'Die Theologie Karl Holls im Spiegel des antiken und reformatorischen Christentums', *Arbeiten zur Kirchengeschichte* 40 (1968) 276 ff.

[2] Oswald Gottlob Schmidt, *Luthers Bekanntschaft mit den alten Classikern* (1883).

[3] Luther, *Fabeln* hrsg. von E. Thiele (1888, 2. Aufl. 1911) pp. 17 f. Thiele published Luther's autograph, found by R. Reitzenstein in the Vatican library; see also O. G. Schmidt (above, n. 2) p. 59, O. Crusius, 'Aus der Geschichte der Fabel' in *Das Buch der Fabeln* von C. H. Kleukens (2. Aufl. 1920) p. xxviii.

[4] See above, p. 88. [5] See above, p. 88.

Erasmus, he said in lectures and writings[1] of the following years, leads away from barbarism to humanity and morality, but no further. Melanchthon demonstrated the weakness of philosophy in his edition of Aristophanes' *Clouds* in 1520; at the same time he argued that the attempt of the medieval Scholastic school to change theology into philosophy had failed completely. In his opinion ancient tradition and philosophy provided useful learned material, mental exercise, and moral teaching, but no help for the understanding of the Bible; faith alone was the way to Christian truth.

Convinced by Luther that the Scriptures alone were the norm (without the tradition of the Church) and that St. Paul alone was the right introduction to the Scriptures, Melanchthon had still to explain why this new personal religious experience and discovery of Luther was really true and binding upon everybody. He tried to give the proof in a compendium *Loci communes* (1521), which originated from lectures on the Epistle to the Romans;[2] despite its title this is not just a collection of relevant τόποι, but a real system of fundamental principles. What he had learnt of ancient rhetoric in his humanistic years, he now applied to Luther's theology. The ancient genre of 'loci' had been revived by Valla in his *Dialecticae disputationes*, by Agricola, and above all by Erasmus in his *Ratio verae theologiae*, where he recommended the use of 'loci theologici' to harmonize the whole world of the Bible and so to facilitate its understanding. But Melanchthon's intention was to build up a true system of fundamental principles, sin and grace, law and faith, and so on, and to give a logical proof of their general validity. Luther himself was astonished at this clever *Graeculus*, and Erasmus called him 'ipso Luthero lutheranior',[3] apparently meaning that he went further than Luther as a strict dogmatist. If the understanding of the Scriptures was left to the religious experience of the individual reader endless confusion might arise; but Melanchthon had now argued that Luther's interpretation was in harmony with principles existing in the Scriptures. With the *Loci communes*[4] Melanchthon laid the foundation of a new branch of literature; his system, the Protestant hermeneutic, a detailed

---

[1] 'De Erasmo et Luthero elogion', 'Ratio discendi', *Corp. Reform.* xx 701 ff.

[2] 'Opera omnia', *Corpus Reformatorum* 1–28 (1834–60) and *Supplementa Melanchthoniana* ii 1 'Philologische Schriften' (1911, reprinted 1968). Melanchthons *Werke* in Auswahl, hrsg. von R. Stupperich ii 1 (1952) 'Loci'; cf. W. Maurer, 'Melanchthon-Studien', *Schriften des Vereins für Reformationsgeschichte* Nr. 181, Jahrgang 70 (1964), esp. pp. 103–36, a quiet and just appreciation of the relation of Melanchthon to Erasmus. See also W. Maurer, *Der junge Melanchthon zwischen Humanismus und Reformation* i (1967) *passim*.

[3] Erasm. *Ep.* 2911.26.

[4] Melanchthon, *Die Loci communes in ihrer Urgestalt*, ed. Th. Kolde (4. Aufl. 1925).

theory of interpretation, was refined and enlarged through the sixteenth, seventeenth, and eighteenth centuries. The *Clavis aurea Scripturae sacrae* of Flacius Illyricus (1567)[1] was the most influential book in this line. The hermeneutic became a part of a new religious theoretical literature, which we may call the Protestant scholasticism.

Did this new theological hermeneutic of the sixteenth century have a significant effect upon classical scholarship? Modern philosophical investigations seem to have favoured that assumption.[2] But, as we have seen, philological interpretation existed from its beginnings as a practice, though not without methodical reflection; there was never a general theory of interpretation, until in the nineteenth century A. Böckh[3] borrowed a sort of hermeneutic from one of his teachers, the theologian Schleiermacher.[4] Böckh's theory, however, as we shall learn, did not influence either his own practice or that of his many pupils. It may have been just that early codification of strict rules by Melanchthon and Flacius Illyricus which prevented the Lutheran exegesis of the Bible from influencing classical scholarship. True scholarship continued to prosper in the Erasmian tradition.[5]

Melanchthon's gift was not to advance scholarship[6] by his own writings, but to procure by his lectures a proper place for the new learning in the universities.[7] In contrast to Erasmus and Reuchlin, he was a very active university teacher all his life; after 1518 he became the magnetic attraction of the university of Wittenberg, whither people flocked to his lectures from all parts of Germany. A Greek scholar from his early youth, he made Greek studies flourish in his university; they had been introduced into Wittenberg from Erfurt in the year of the foundation of the new university in 1502/3. It was not the case that the study of Greek suddenly started together with the Reformation; it

---

[1] G. Moldaenke, 'Matthias Flacius Illyricus', *Forschungen zur Kirchen- und Geistesgeschichte* 9 (1936).

[2] W. Dilthey, 'Die Entstehung der Hermeneutik', *Philosophische Abhandlungen Christoph Sigwart gewidmet* (1900) pp. 185–202 = *Gesammelte Schriften* v (1924) 317 ff. and 426 f.; cf. G. Ebeling, 'Hermeneutik', *Die Religion in Geschichte und Gegenwart* III[3] (1959) 242–62.

[3] *Vorlesungen über Enzyclopaedie und Methodologie der philologischen Wissenschaften*, printed after his death 1877 and 1886.

[4] He defined it correctly as 'Kunstlehre des Verstehens'; it should not be identified with interpretation, as is so often done.

[5] See above, p. 82.

[6] Cf. above, p. 88. Only recently we were reminded by W. Beneszewicz, 'Melanchthoniana', *SB Bayer. Akademie, Philos.-hist. Abt.*, Jg. 1934, Heft 7, that Melanchthon was the first to draw attention to the so-called 'Apostolic canons' and that his small first edition of the Greek text (1521) for his students became the basis for later more scholarly editions.

[7] On their conservative attitude see above, p. 89.

advanced by a rather slow and complicated process.[1] After Reuchlin no one did more to further it than Melanchthon; but he seems also to have been responsible for a certain classroom atmosphere which became characteristic of the later German humanism in contrast to that of France and England.

The traditional label 'Praeceptor Germaniae'[2] for Melanchthon is entirely justified. The effect of his teaching on the Protestant schools was enormous, and Wittenberg continued to be the central university where most of the teachers were trained. The outstanding pupils of Melanchthon were Joachim Camerarius (1500–74), a future headmaster of the high school in Nuremberg, and Hieronymus Wolf (1516–80), who was to hold a similar position at Augsburg. They came under the progressive influence of the new scholarship in France[3] and left Melanchthon's other friends and pupils far behind them; but with these younger scholars we are already beyond the age of Erasmus.

There was little to be said about the slow and late start of humanism and scholarship in Spain;[4] but we did notice an outstanding work of serious scholarship produced there in the early sixteenth century, the first Polyglot Bible. It was, however, not able to compete with Erasmus's Greek New Testament[5] which became more and more popular everywhere, even in Spain. This led to bitter attacks upon Erasmus's work by one of the collaborators of the Polyglot Bible, Jacobus Lopis Stunica (Zuñiga), and Erasmus himself was not slow in launching his counterattacks.[6] Inevitably there were errors and weaknesses on both sides, and a scrupulous re-examination of the reliability of each biblical text would hardly be worthwhile.[7]

A more competent and sensible Spanish opponent of Erasmus than Stunica was Juan Ginéz Sepulveda (1491–1572).[8] After a few years in Cordova and Alcalà, his scholarly education took place at Bologna and in other cities of Italy, where he remained for nearly twenty years. There is even a puzzling 'Bentley touch' in one of his last letters to

---

[1] See *Gymnasium* 71 (1964) 201, and Beneszewicz, 'Melanchthoniana' (above, p. 93 n. 6) pp. 16 ff. on printers of Greek texts in Wittenberg.

[2] K. Hartfelder, 'Philipp Melanchthon als praeceptor Germaniae', *Monumenta Germaniae paedagogica* 7 (1889).

[3] See below, p. 139.   [4] See above, pp. 61 and 65.   [5] See above, pp. 76 f.

[6] See Erasm. *Ep.* XII (1958) Indices pp. 172 f. s.v. Stunica and p. 17 Erasmus's writings with all the titles of the mutual polemics; cf. also Bataillon (below, p. 95 n. 4) pp. 98 ff.

[7] See F. G. Kenyon, *Handbook of the Textual Criticism of the New Testament* (2nd ed. 1926) pp. 267 ff.; *The Cambridge History of the Bible. The West from the Reformation to the present day* (1963) pp. 56 ff.; cf. Bataillon (below, p. 95 n. 4) pp. 101 f.

[8] Erasm. *Ep.* 2637 with Allen's introduction on Sepulveda's life and works. *Opera*, 4 vols. (Madrid 1780).

Erasmus: 'mihi tamen certissima ratio semper erit pluris quam omnes enarratores vel interpretes.'[1] Sepulveda was the first to collate the fourth-century Vatican manuscript of the Bible, later called Vaticanus B. Though it was in the Vatican before 1481 and known to Stunica, it was not used for the Complutensian Polyglot.[2] Sepulveda compiled a list of 365 variant readings in B, as he wrote to Erasmus in 1533,[3] but whether he sent him the list, we cannot tell; in any case, Erasmus did not realize the importance of the codex and was mistaken about the relation of its readings to the Vulgate.

In spite of these controversies on particular questions, a general 'Erasmianism' gained a central place in Spanish cultural life, at least for a few decades.[4] The fundamental ideas of the 'Philosophia Christi' were effective, less in the field of classics itself than in those of theology and Roman law. There the outstanding representative was Don Antonio Agustín[5] (1517–86) who studied under Alciato[6] at Bologna where Antonio de Lebrija had studied before him. His collation and editing of the famous Florentine manuscript of the *Digests*[7] was so careful that even after 300 years it earned him the applause and admiration of Mommsen.[8] Agustín used his years in Italy also for editing Varro *De lingua Latina* (1554) and Festus (1559) and for the intensive study of inscriptions and antiquities. Subsequently he returned to Spain to become bishop of Lerida and finally archbishop of Tarragona. His library, especially because of its Greek manuscripts, is one of the chief treasures of the library in the monastery of the Escorial founded by Philip II between 1566 and 1587.[9]

On the whole, however, 'Erasmianism' did not flourish under Philip's regime; we have to look back once more to the time of Charles V to find a Spaniard who became one of the most devoted Erasmians in Europe,

---

[1] *Ep.* 2938.27 ff.

[2] F. Delitzsch, 'Studien zur Entstehungsgeschichte der Polyglottenbibel des Cardinals Ximenes', [*Programm*] *zur Feier des Reformationsfestes . . . der Universität Leipzig* (1871) pp. 13 ff., seems to have been the first to try to unravel the complicated and often confused history of the sources of the Complutensis, see also ibid. (1886).

[3] *Ep.* 2873 with Allen's notes; cf. *Ep.* 2905.

[4] M. Bataillon, *Érasme et l'Espagne* (1937, 2nd Spanish ed. corrected and augmented 1966), the model of an exhaustive and sympathetic monograph. The references are to the French original of 1937. Lluis Nicolau d'Olwer, 'Greek Scholarship in Spain', in James Kleon Demetrius, *Greek Scholarship in Spain and Latin America* (1965) pp. 15–22, followed by bibliographies pp. 33–131.

[5] *Opera*, 8 vols. (Lucca 1765). F. de Zulueta, *Don Antonio Agustín*, Glasgow University Publications 51 (1939).

[6] See above, p. 86.   [7] See above, p. 43.

[8] *Digesta* ed. Th. Mommsen (1870) vol. I, pp. xvi f.

[9] E. Jacobs, *Zentralblatt für Bibliothekswesen* 25 (1908) 19 ff.

Juan Luis Vivès[1] (1492–1540). In contrast to all the other Spaniards so far encountered, he did not visit Italy, but went first to Paris and then to the southern Netherlands under Spanish dominion. In the university of Louvain the writings of Erasmus, and possibly personal acquaintance with him, converted Vivès from the Scholastic philosophy of his native country and of Paris to Erasmian humanism. His several educational books, especially *De disciplinis* (1531), and his edition of *De civitate dei* bear witness to this conversion.

Clearly it was Italian humanism that inspired the Spanish scholars, with the one exception of Vivès. When Erasmus became suspect of heresy about 1550, the anti-humanistic forces suppressed the Erasmian and similar tendencies. His followers were first threatened by the Inquisition and then from 1557/8 onwards subjected to actual persecution. The glorious rise of scholarship in France thus came too late to have any effect upon Spain.

[1] *Opera* (Valencia 1782–90, reprinted 1964); cf. Bataillon (above, p. 95 n. 4) pp. 655 ff. and *passim*.

PART THREE

FROM THE FRENCH RENAISSANCE TO THE GERMAN NEOHELLENISM

IX

HUMANISTS AND SCHOLARS OF THE FRENCH RENAISSANCE

MANY paths radiated from the centres of Italian humanism to the western and northern countries, but the main road led to Paris.[1] France was the scene of the most lively activity, and in the sixteenth century it surpassed all other countries in Greek scholarship. This spectacular achievement, although it was the culmination of a slow process of evolution from Petrarch's time onwards,[2] went far beyond the promise of earlier developments, and its origin has not yet been convincingly explained. We shall find its immediate cause, not in social or political influences, but in the inspiring force of *one* individual *poeta doctus*.[3] Yet it remains true that French cultural life was strongly determined by politics. A consolidated monarchy had existed since the fifteenth century with a governing class of nobility and clergy around the king. Beside the more conservative university of Paris, the royal court formed a second cultural centre, which promoted the conquest of the treasures of antiquity as an object of national ambition.

At the court of Louis XII Claude de Seyssel (1450–1520) was indefatigable in translating ancient texts, both Latin and Greek, into French for the glory of the nation.[4] As he said in the preface to his Justinus (composed in 1509, but not printed before 1559), 'the Romans in conquering Greek literature made their own language powerful and prepared a medium for world domination; we have to translate Latin and to create a mighty French literature.' Under Francis I[5] this

---

[1] For the French Renaissance in general and its relation to Italy see the recent studies of Franco Simone, especially *Il Rinascimento Francese, Studi e ricerche* (1961), and *Umanesimo, Rinascimento, Barocco in Francia* (1968). A. Buck, *Die humanistische Tradition in der Romania* (1968), esp. pp. 133 ff. 'Humanismus und Wissenschaften'; cf. J. von Stackelberg, *Gnomon* 42 (1970) 424 ff.

[2] See above, pp. 60 f., 65.   [3] See above, pp. 103 ff.

[4] P. H. Larwill, *La théorie de la traduction au début de la Renaissance* (Diss. München 1934) pp. 38 ff. Following F. Brunet, Larwill stressed Seyssel's effect on Du Bellay (1549); but he might have noticed the strong influence of Valla's spirit and eloquence on Seyssel. In translating Greek historians he had the help of Janus Lascaris, see below, p. 102.

[5] A. Tilley, *Studies in the French Renaissance* (1922) pp. 123 ff. 'Humanism under Francis I'.

ambitious cultural nationalism was transferred from books to life: the Italian Renaissance as a whole was to be surpassed by France. In the end the French believed that they had surpassed not only the modern Italians, but the ancients themselves; 'la querelle des anciens et modernes' at the end of the seventeenth century[1] illustrates this bold self-assurance. Scholarship itself, however, derived no great profit from this nationalistic line of thought.

In contrast to it the more international spirit was represented in Paris by the philosopher Lefèvre d'Étaples (Faber Stapulensis, c. 1455–1536). Influenced by the Florentine Platonists and by personal contact with Erasmus after 1511,[2] he tried to restore a genuinely philosophical education. Unlike the late Scholastic philosophers he did not rely on translations of Aristotle, but used the original Greek text, and he also translated and explained parts of the Scriptures, especially St. Paul's Epistles. His tolerant and cautious humanism was far removed from the spirit of the radical religious reformer Jean Calvin (1509–64). Nevertheless Calvin was in much closer contact with the humanistic movement than Luther. He began his career in 1532 with the publication of a commentary on Seneca's *De clementia*.[3] As a pupil of Budé, of the first two Royal Readers in Greek at the Collège de France,[4] and of Andrea Alciato, who lectured on Roman law at Bourges, he was thoroughly trained in classical scholarship and Roman law for his later biblical exegesis, and was familiar with all the philosophical, theological, and juridical elements essential to his new dogmatic, the *Institutio religionis Christianae* (1536).[5] His Latin style was concise, pertinent, impressive; his basic ideas were certainly not humanistic in the sense of Petrarch or Erasmus, but a rational desire for knowledge and for independence, together with an energetic activity in the consciousness of being the 'elect of God' created new impulses to research.

Looking at French cultural life as a whole in the first half of the sixteenth century, we see a sometimes bewildering picture of various

---

[1] See below, p. 134.

[2] P. S. Allen on Erasm. *Ep.* 315 and *Ep.* xii (1958) Indices, pp. 17 and 90; cf. P. Mestwerdt, *Die Anfänge des Erasmus* (1917) pp. 323 ff. and A. Renaudet, *Préréforme et humanisme à Paris 1494–1517* (2nd ed. Paris 1953) esp. p. 703.

[3] Calvin's commentary on Seneca's *De clementia*. With introduction, translation, and notes by F. L. Battles and A. M. Hugo, 1969 (Renaissance Text Series of the Renaissance Society of America).

[4] See below, p. 102.

[5] The first Latin edition and the first French translation with a dedicatory letter to Francis I were published in 1536, see Calvin, *Opera selecta* ed. P. Barth and G. Niesel iii (1957) vi ff.— On Calvin's 'humanism' see Jean Boisset, 'Sagesse et sainteté dans la pensée de Jean Calvin', *Bibliothèque de l'école des hautes études*, section des sciences réligieuses 71 (1959) 315 ff.

colours.[1] The first leading figure to emerge in the field of classical scholarship was Guillaume Budé (1468–1540). Though a friend of Erasmus (whose exact contemporary he was)[2] and of Reuchlin and in lively correspondence with both of them, he was a man of different type. One might call him a 'bourgeois Parisien', open to the details of daily life and interested in economics, politics, and law, a man who had been in the active service of his kings[3] before he retired to a life of pure scholarship in his comfortable house in town and in two cottages in the country.[4]

With his *Annotationes ad Pandectas*[5] (1508) Budé left his predecessors, Valla, Politian, and Zasius, far behind him and prepared the way for the great jurists of the next generation like Cuiacius; for he was not only well versed in the ancient languages, but possessed a much greater knowledge of the subject-matter, which enabled him to interpret particularly difficult passages of the pandects in their proper context. His treatise *De Asse eiusque partibus* (1514), based on the careful research of nine years, became the definitive textbook for the study of Roman coinage and metrology. So great was the attraction for his fellow-countrymen of the material side of ancient life, and especially of social life, that it reached its tenth edition in the course of twenty years. That he had acquired a remarkable knowledge of Greek was shown in his later books, *Commentarii Linguae Graecae* (1529) and *De Transitu Hellenismi ad Christianismum* (1535). The *Commentarii* were a preparatory work for a comprehensive Greek dictionary, a *Thesaurus Linguae Graecae*,[6] the most original and striking parts of which concerned the Greek and Roman legal terminology. He certainly deserved to be called by the greatest and most competent critic 'le plus grand Grec de l'Europe'.[7] It is significant that he gave his general account of classical scholarship the title *De Philologia* (1532).[8] He was indeed a φιλόλογος in the sense of Eratosthenes, who had been the first to claim this *cognomen* for himself; it referred, according to Suetonius, to persons familiar with the various branches of knowledge or even with the whole of the λόγος.[9]

---

[1] Cf. Carl J. Burckhardt, *Vier historische Betrachtungen* (1953) p. 10.

[2] See Erasm. *Ep.* vol. XII Indices, pp. 59 f.   [3] See above, p. 65.

[4] For some amusing anecdotes see Sandys II 171 f.

[5] *Omnia Opera* (Basle 1557, repr. 1967); L. Delaruelle, *Répertoire analytique et chronologique de la correspondence de G. Budé* (1907). Bibliographical references in A. Buck's introduction to *De Philologia*, see below, n. 8; on legal studies see above, p. 86 with notes.

[6] See below, p. 110.

[7] *Scaligerana*, editio alphabetica Coloniensis (1595) p. 72.

[8] G. Budaeus, *De Philologia. De studio litterarum*. Reprinted with an introduction by A. Buck (Stuttgart 1964).

[9] See *History* [1] 158 f., 170.

Encyclopedic knowledge, not eloquence, leads to true human culture; that was Budé's conviction. In *De Transitu* he was moving away from the Erasmian concept that Hellenism, especially Greek philosophy, prepared the way for Christianity, and beginning to lay more stress on the difference between Hellenism and Christianity. It may be that in those later writings he was somewhat under the influence of Calvin.[1]

One of Budé's greatest achievements, perhaps even his greatest, was to induce his king, Francis I (1515–47), to found the Collège Royal (later Collège de France) for promoting the new learning. In this enterprise Budé enjoyed the efficient help of Janus Lascaris, a Greek scholar who paid three visits to France between 1495 and 1534 as the guest and helper of three kings, Charles VIII, Louis XII, and Francis I. Previously he had been active for a considerable time in Italy;[2] and between his first and second visits to Paris he belonged to that philhellenic society in Venice[3] which was so successful in furthering the transmission of Greek language and literature to the West. Lascaris taught his French friends the Greek language, provided them with recently printed Greek texts (he was himself a competent editor of ancient authors), and assisted those who tried to translate from Greek into French. Budé, in his *Commentarii Linguae Graecae* (1529), described the proposed royal college as a new Μουσεῖον. This unmistakable reference to the famous foundation of Ptolemy I in Alexandria[4] served also to indicate the difference between the college and the philosophical Academy[5] of the Medici in Florence; it was meant to be a free community of poets and scholars protected by the Muses. The name 'Pléiade', probably given by Ronsard in 1563 to a smaller circle of seven poets and friends who were members of the Museum likewise refers to Alexandria.[6] After thirteen years of preliminary discussions the Collège Royal opened in 1530. Two *lecteurs royaux* for Greek (not for Latin) were immediately appointed, Pierre Danès and Jacques Toussain; both were pupils of Lascaris, Toussain also a pupil of Budé and a friend of Erasmus. Their lectures attracted a great many listeners, amongst them some of the highest rank: Calvin, Rabelais, Ignatius of Loyola, Amyot, Ronsard, Henri Étienne, Francois de Sales.

One of Toussain's pupils was Jean Dorat (1508–88).[7] Dorat's natural

---

[1] J. Bohatec, *Budé und Calvin. Studien zur Gedankenwelt des französischen Frühhumanismus* (1950) esp. pp. 82 ff. See also Buck's introduction to *De Philologia* (above p. 101 n. 5) p. 22 with further references.

[2] See above, pp. 48, 61.   [3] See above, p. 56.   [4] *History* [1] 98, 119.
[5] See above, pp. 57 f.   [6] *History* [1] 119.
[7] The important biographical details are preserved only by J. A. de Thou, *Mémoires*. First

gift enabled him to write elegant verses in three languages, Greek,
Latin, and French; deservedly he became *poeta regius*—though we had
better not record the astronomical figures of his verse production.
What concerns us here is his sincere enthusiasm for the great poetry of
the past and his ability to instil his own love of it into the minds of his
friends and pupils and to inspire succeeding generations. In his scholarly
work he concentrated on Greek poetry, interpreting it and correcting
corrupt texts, a task made possible by the preparatory work carried out
jointly by Greek and French scholars in Paris. As a young man Dorat
had been acquainted with Budé and had attended the Greek lectures of
Germain de Brie (who, like Toussain, was a pupil of Lascaris and a
correspondent of Erasmus).[2] He started his career about 1544 as a
private teacher in the house of Lazare de Baif instructing Lazare's son
Jacques Antoine and Pierre Ronsard,[3] both of them future poets, in
Greek and Latin, but especially Greek. Then from 1547 onwards he
taught at the Collège de Coqueret, which belonged to the university of
Paris, and was for four unhappy years tutor to royal princes, a position
similar to that held by a number of his Alexandrian predecessors in the
third and second centuries B.C.[4] Relatively late, in 1556, he was
appointed *lecteur royal* beside Turnèbe at the Collège Royal, thus earning
the title of 'interpres regius', which he retained even after 1567, when he
voluntarily returned to teaching a circle of selected private pupils. We
may apply to him the designation given to Philitas about 300 B.C.,

French edition (Amsterdam 1711) p. 6; see also P. de Nolhac (this note below) p. 45.1.
'Dichter und Philologen im französischen Humanismus', *Antike und Abendland* 7 (1958) 73–83.
This short paper is based on more extensive lectures, first delivered in May 1956 in Basle and
Zurich. It was a lucky coincidence that in the same year in which I delivered these lectures
on the literary revival of Greek in France Dora and Erwin Panofsky, *Pandora's Box*, Bollingen
Series 52 (1956, rev. 2nd ed. 1962) 55 ff., pointed out that there was in France a Greek revival
in the arts unparalleled elsewhere. I owe the knowledge of this article to the kindness of my
old friends, the authors. It was only in preparing the lectures that I began to see clearly the
decisive part played by Dorat at this turning-point in the history of scholarship in France.
Mark Pattison, *Essays* 1 (1889, repr. 1965) 206 ff. and his notes in MS. Pattison 79–93,
Bodleian library, cf. below, p. 119 n. 7 (on p. 120). Pierre de Nolhac, *Ronsard et l'humanisme*
(1921) *passim*; cf. also Tilley (above, p. 99 n. 5) pp. 219 ff. 'Dorat and the Pleiade' and H.
Chamard, *Histoire de la Pléiade*, 4 vols. (1939–41). Wilamowitz, *Geschichte der Philologie* (1921)
p. 25, put Dorat at the end of the French Renaissance and so obscured his historical position;
he dismissed him in a single sentence blaming Gottfried Hermann for having exaggerated the
importance of Dorat's Aeschylean emendations (see below, p. 104 n. 2). Wilamowitz, of
course, had no sympathy for a scholar who only lectured and taught, but did not write down
and publish his own studies. The dates of Dorat's life are confused in Sandys II 185.
  [2] Twenty letters of Germain de Brie are preserved in the correspondence of Erasmus, see
Ep. XII, Indices, p. 3 (Brixius).
  [3] See Ronsard, *Oeuvres complètes*, ed. P. Laumonier XVI (1950) 5, 'la naïve facilité d'Homère
. . . la curieuse diligence de Virgile', the editor refers to Quintil. x 86.
  [4] See *History* [i] 92, 154.

ποιητὴς ἄμα καὶ κριτικός,[1] but we must confess that his scholarship far surpassed his poetry. In a humorous way he once alluded to the importance of his being a poet: when at the age of 78 he lost his wife and married a girl of 19, he said that as a *poeta* he was entitled to *licentia poetica*.

As an 'interpres' Dorat was the first to explain all the Greek dramatists, the tragic poets as well as Aristophanes. Starting with Euripides, he joined Ronsard and Baif in helping Tissard to translate the tragedies into French. Then he turned to Aeschylus, beginning with the *Prometheus* and intensifying his efforts after the publication of Petrus Victorius's edition in 1557; it was his success as a critic of the text of Aeschylus that won him immortality.[2] A copy of a late lecture on Sophocles is preserved by chance. Of the lyric poets his interpretations of Pindar made a great impression as early as the forties. His treatment of the late Anacreontics (believed to be genuine in the sixteenth century) anticipated Henri Étienne's *editio princeps*. His lectures on the *Iliad* and *Odyssey*[3] at the Collège de Coqueret and the Collège Royal became very famous; unlike J. C. Scaliger he never doubted the superiority of Homer. From the early epic, lyric, and dramatic poets he moved on to the later Greek poets. Manuscript notes reveal how he explained to his pupils the mass of Greek epigrams printed by Lascaris. The *hymns* of Callimachus, rarely understood at that time in spite of Politian's[4] efforts, were among his favourites, and so were Callimachus' contemporaries, Aratus, Theocritus, Apollonius Rhodius, and even Lycophron. His interest extended from Nicander, Moschus, Bion, and Oppian down to the *Dionysiaca* of Nonnus, which he had read and explained a long time before they appeared in print in 1569. It is highly probable that he introduced his pupils to three 'poetical' dialogues of Plato; it is quite certain that he opened the way for them to all that body of Greek poetry which was previously almost unknown in France and the other countries of Western Europe.

Dorat fascinated his large audiences, which included (since he was a

---

[1] See *History* [1] 89.

[2] G. Hermann on Aesch. *Ag.* 1396: 'ille omnium qui Aeschylum attingunt princeps Auratus'; cf. E. Fraenkel in his commentary on *Agamemnon* 1 p. 35. See also P. de Nolhac (above, p. 102 n. 7 [on p. 103]) Index, Éschyle, and below, p. 106.

[3] G. Canter, . . . Novarum lectionum libri octo (Antwerp 1571, 3. ed.) 333–7. On the allegorical interpretation of Dorat and his contemporaries N. Hepp, 'Homere en France au XVIe siècle', *Atti della Accademia delle Scienze di Torino, II. Classe di Scienze Morali, Storiche e Filologiche* 96 (1961/2) 428 ff. and *Homère en France au XVIIe siecle* (1968) with A. Buck's review, *Gnomon* 45 (1973) 291 ff.

[4] See above, p. 45.

scholar poet himself) not only future scholars, but also the flower of young poets. Greek was a revelation to them and incited them, even more than Latin to imitation and rivalry. No one has better expressed the effect of Dorat's teaching than Étienne Pasquier, a friend of Montaigne, in a Catullan poem, beginning with 'Auratus meus ille quem videtis' and ending with the pointed lines: 'scribunt carmina ceteri poetae, / summos at facit unus hic poetas.'[1] That this must be an allusion to the *poeta doctus* Valerius Cato, the head of the circle of the *poetae novi* of which Catullus was the greatest, shows the warmth of the compliment. Ronsard[2] said the same in simple words: 'Dorat m'apprit la poésie.' A contemporary source preserves the anecdote that when Dorat read the complete *Prometheus*[3] to his pupil in French the young Ronsard, deeply moved, exclaimed: 'Et quoy, mon maistre, m'avez vous caché si long temps ces richesses?' Aeschylean images abound in Ronsard's lyric poetry. But he found a still higher inspiration in Pindar,[4] when Dorat had read and explained to him the four books of the Epinicia: 'les saintes conceptions de Pindare' was Ronsard's expression of his immediate reaction. At Dorat's suggestion he even tried in his four books of *Odes* (1550) to introduce the triadic structure of Pindar's poems into French poetry, in place of the simpler forms of Horatian and Catullan lyrics. As Dorat had impressed the original connection of lyric poems with music on the mind of his pupils, Ronsard was proud when his odes were set to music by Orlando di Lasso and others. J.-A. de Baif,[5] once Dorat's pupil in his father's house together with Ronsard, tried hard to write quantitative French verses and to have set them to music.[6] We even have a setting of the Greek text of Pindar *O*. I that preserves the original metre;[7] this could only have been produced in France at that time and under Dorat's influence, for otherwise composers remained faithful to the Latin tradition.

It is evident that there was a continuous interplay between poets and scholars, most lively in the circle of the Pléiade.[8] One can hardly say whether classical studies owed more to contemporary poetry or poetry

[1] P. de Nolhac (above, p. 102 n. 7 [on p. 103]) quotes the whole poem from Pasquier's *Oeuvres* (p. 57), but it escaped him that these concluding lines are borrowed from the anonymous hendecasyllables preserved in Suetonius, *De grammaticis* c. 11 (*Fragmenta poetarum Latinorum*, ed. Morel, 1927, p. 83): 'Cato grammaticus, Latina Siren, qui solus legit ac facit poetas.'

[2] P. de Nolhac (above, p. 102 n. 7 [on p. 103]) p. 53. [3] Cf. below, p. 106.
[4] P. de Nolhac pp. 49 ff. and *passim*. [5] Cf. above, p. 104.
[6] On metrical composition as invented by Celtis see above, pp. 63 f.
[7] R. Wagner, *Philol.* 91 (1936) 170 and in general Chamard, *La Pléiade* IV 133 ff. 'l'alliance de la poésie et de la musique.'
[8] See above, p. 102.

to scholarship. So far we have emphasized Dorat's effect on the poets ('facit poetas', 'm'apprit la poésie'); but one of his pupils, Marc-Antoine de Muret,[1] who wrote a commentary on Ronsard's *Amours* of 1552, called him 'omnium eruditorum magistrum'. From the context it is clear that 'eruditi' here means learned men, or scholars. Dorat's great success as a 'magister' is understandable: he took immense trouble to collect and examine the variant readings in Greek manuscripts and to recognize the special characteristics of poetic language and style. The oratorical delivery of the results of his researches filled his audiences with enthusiasm. He seems to have lectured most frequently on Homer; his brilliant young Dutch pupil Willem Canter, who called him 'unicum et optimum Homeri interpretem', has preserved excerpts from a lecture on the *Odyssey* with strange allegorical interpretations.[2] The epithets 'unique and best' had better be reserved for Dorat as interpreter of Aeschylus.[3]

Dorat did not trouble to have his emendations and interpretations published. He once seriously considered printing a Latin translation of Pindar, with a dedication of each of the forty-six Epinicia to individual friends, but he never did so. Although bibliographies and handbooks[4] usually list an edition of Aeschylus' *Prometheus*[5] by Dorat (1549), no one has ever been able to prove its existence.[6] This is characteristic: Dorat exercised his stupendous influence not through writing, but through the spoken word. What he said was immediately taken up by his pupils, and through them passed on as oral tradition to at least two further generations.

Though Dorat did not care to have his own researches printed, he strongly encouraged the general development of the printing of classical Greek books in France. The most important instruments for research, the full collection of Greek texts and the complete dictionary, were produced in a few decades. We should not look only at the dates of publication from the sixties to the eighties of the sixteenth century, but should be conscious also of the creative impulses in the middle of the century. For instance, we can find traces of Dorat's teaching in the innumerable editions of that most prolific scholar, Henri Étienne, Robert's son. In his complete collection of Greek epic poets he quoted unpublished readings of Dorat with moving devotion. He did the same in his edition of Aeschylus and in the *editio princeps* of the Anacreontics

---

[1] P. de Nolhac pp. 92 ff.   [2] See above, p. 104 n. 3.   [3] See above, p. 104 n. 2.
[4] Sandys II 187.   [5] Cf. above, pp. 104 and 105.
[6] P. de Nolhac (above, p. 102 n. 7) p. 44 n. 2 'purement imaginaire'.

(1554), the *codex unicus* of which he had discovered himself in Italy some time before 1549 and made accessible to Dorat and other friends. It is still possible to follow the various ways by which Dorat's conjectures were saved for posterity;[1] usually his own pupils put down those that they had heard, and then a fair number of them were taken over into marginal notes of printed editions.

The king who had founded the Collège Royal also founded the Presse Royale and had a magnificent Greek type executed by the best French engraver, Claude Garamond; it was cast at the expense of the treasury in 1541 and called 'typi regii'.[2] Printers of Greek all over Europe treated the 'Royal Greeks' as a model for more than two centuries. The first printer to use them was Robert Étienne.[3] His father, Henri I, had founded a printing office in Paris at the beginning of the century and his eldest son, Henri II, kept the press in feverish activity till his death in 1598. Robert was at once author, printer, reader, and publisher. He made his reputation with a Latin dictionary which, starting as a revision of the dictionary of Calepinus, a rather poor Italian schoolbook, had grown by 1543 into the *Latinae Linguae Thesaurus* in three folio volumes. It had to satisfy the demands of more than two centuries until Forcellini's *Totius Latinitatis Lexicon* was published in Italy in 1771, again the work of a single man. Indeed it still holds its place, after five revisions during the nineteenth century.[4]

Almost simultaneously with Stephanus's Latin *Thesaurus* there appeared the *Commentarii Linguae Latinae* of Étienne Dolet (1509–46),[5] an exposition for the most part of Ciceronian usage; they were followed by his book of Ciceronian phrases. Then turning to Greek, he was the first to translate Platonic dialogues into French. Charged with 'heresy', he died as a 'martyr'.

[1] A. Tilley (above, p. 99 n. 5) pp. 220 ff.

[2] There were three alphabets in different sizes, each alphabet having about 400 characters, because of the many ligatures; a simple type with 40 characters without ligatures was introduced not earlier than the end of the seventeenth century; see V. Scholderer, *Greek Printing Types* (1927).

[3] Elizabeth Armstrong, *Robert Estienne, Royal Printer*. An historical study of the elder Stephanus (Cambridge 1954) p. 33, fig. 8 Royal Greek types. A comprehensive monograph, well documented and illustrated.

[4] Robert's *Thesaurus* and Forcellini's *Lexicon* are going to be superseded by one of the greatest collective enterprises of the last century, the new *Thesaurus Linguae Latinae*; after long and tiresome preparations the first fascicles were printed in 1900, and it is estimated that now in the seventies about half of the *Thesaurus* has been published. See 'Klassische Philologie' in *Geist und Gestalt*, Biographische Beiträge zur Geschichte der Bayerischen Akademie der Wiss. 1 (1959) 123 on Karl Halm's preliminary plan, 126 ff. on Eduard Wölfflin's new foundation and on his many collaborators.

[5] R. C. Christie, *Étienne Dolet* (1880, 2nd ed. 1899, repr. 1964) pp. 234–62 on *Commentaria* and *Formulae*.

In the year after the completion of the first Latin *Thesaurus* Robertus Stephanus turned to the printing of Greek books.[1] He began in 1544 with the *editio princeps* of Eusebius' ecclesiastical history, following it with seven other first editions. But his own special desire was to spread the knowledge of the Scriptures; and from 1545 onwards he published several editions. His first production was the folio edition of the Greek Testament in 1550, in point of beauty of execution still the most perfect edition ever printed.[2] The text was that of Erasmus's third edition of 1535,[3] but variant readings of fifteen manuscripts were added in the margins. As the King's printer, he had been engaged in a continuous and damaging feud with the old university of Paris. Now, because of renewed difficulties, he removed the chief part of his press to Geneva, and there in 1551 he made an open profession of the reformed faith. While his second son Robert remained in the old house in Paris as a Catholic, his eldest son Henri became his successor in Geneva, the two Stephanian presses continuing to exist without any hostility. The Geneva edition of the Greek Testament of 1551 is remarkable in being the first in which the text was divided into verses.[4] The division of the text of the New Testament into chapters (κεφάλαια) is to be found in manuscripts from the fourth century onwards, and probably had its origin in liturgical use.[5] Robertus Stephanus cut the chapters into shorter sections (τμήματα, *sectiunculae*) and numbered them by figures. We are told that he carried out this operation while travelling on horseback from Paris to Lyons, perhaps during the journey by which he left France for good; he extended the system from his Greek New Testament of 1551 to the Latin Old Testament in his edition of 1556. All Protestant printers adopted this useful innovation, and the printing of Stephanus's verse numbers in the definitive Catholic edition of the Vulgate in 1592 set the pattern for Roman Catholic Bibles also; in this respect at least all denominations are united.[6] The second remarkable success of Robert's Greek Testament was that his text was reprinted by all the presses of Europe, becoming 'textus receptus', as it was

---

[1] See above, p. 106.

[2] *The Cambridge History of the Bible.* [II:] The West from the Reformation to the present day. Ed. by S. L. Greenslade (1963) p. 438.

[3] See above, p. 76.

[4] The printed Latin Bible of Froben (1491) had its pages numbered and marked down the margin by the first letters of the alphabet at equal intervals. We still use this form of reference for Plato, Plutarch, Strabo, and Athenaeus.

[5] B. M. Metzger, *The Text of the New Testament* (²1968) pp. 22 f.

[6] It has been regretted in our time that this traditional division does not conform to the real rhythm of biblical prose—but that would have been to expect too much from Robert's work on horseback or at a resting-place

called in the Elzevier edition of 1633,[1] and was never altered until Lachmann's completely new critical recension of 1831 made a fresh start necessary.

When Robert Étienne died in 1559 in Geneva, the press passed to the eldest of his nine children, Henri (who was born probably in Paris in 1531). Latin was, we may say, his mother tongue; he learnt Greek as a child and at the age of eleven attended the lectures of the great classical scholars at the Collège Royal. The forties of the sixteenth century were just the age of that passionate Grecism in France which we have tried to describe; Henri Étienne was imbued with the deepest love of the Greek language and became incredibly familiar with its idiom. He really thought in Greek and could speak it; to him it was simply not a foreign language at all. In this respect he was, as far as I can see, unique. He was indeed not an ordinary academic grammarian or critic, but a great adventurer in the field of Greek scholarship. He was prolific in publishing texts, amongst them seventy-four Greek texts of which no fewer than eighteen were first editions. On long travels[2] through Europe he collected and collated manuscripts, corrected the Greek text, read the proofs, revised or made a Latin translation, and often added notes and appendices. At his peak he managed to produce about 4,000 pages of Greek texts in a year. His discovery[3] and publication of the *Anacreontea* (1554) was a sensation of the first class and the starting-point for a new branch of modern literature. He made the first collection of the fragments of all the Greek lyric poets in 1560; the third and final edition appeared in 1586. To his edition of the *Anthologia Planudea* in 1566 he added a collection of the numerous epigrams quoted by Greek authors. His Plato (1578) became the standard edition, and we still quote the numbers of its pages. Some of his other editions became the vulgate texts for two centuries or even longer, not always to the advantage of scholarship. For Henri Étienne was neither truly critical nor careful, and had neither sense of poetry nor literary taste. That is the flaw in the work of this passionate genius. His prefaces are full of personal remarks

---

[1] 'Textum habes nunc ab omnibus receptum, in quo nihil immutatum aut corruptum damus.' Preface of the second Leyden edition.

[2] On his relations as printer and book-collector to the family of the Fuggers in Augsburg see below, p. 140.

[3] It is generally assumed that Henricus Stephanus brought in 1549 a manuscript containing the *Anacreontics* from Italy to Paris (so Chamard, *La Pléiade* II 56). But see *Anacreontea* ed. K. Preisendanz (1912) pp. vii ff. following Paul Wolters: Henricus Stephanus, returning from England in 1551, passed through Louvain; there Ioannes Clemens Anglus had a copy of the *Anacreontics* of which Stephanus made a copy which he showed to some of his friends and published in 1554. The original cod. Palatinus plays no part in this story.

and would provide ample material for an extensive authoritative biography.[1]

Besides his Greek editions he produced fifty-eight Latin and three Hebrew texts and thirty books of his own in Latin or French. One of them is of special interest to us, as it was the first modern history of classical scholarship, *De criticis veteribus Graecis et Latinis* (1587). His most popular work was his *Traité préparatif à l'apologie pour Hérodote* (1566), a volume of 600 closely printed pages, which passed through fourteen editions during his lifetime. It is an amusing collection of short stories and anecdotes, much more appreciated in Paris than in Calvinistic Geneva. Next to Greek, as superior to all languages, he was interested in French as superior to the other modern languages, and on the special request of King Henri III he wrote in a fortnight a book on the *Précellence du langage françois*.

Only if we consider his unique gift for language and the fact that in editing texts he was interested above all in the language shall we be able to understand his greatest achievement, the Θησαυρὸς τῆς Ἑλληνικῆς γλώσσης, published in five volumes in 1572. This *Thesaurus Graecae Linguae* was re-edited in another arrangement and with supplements twice in the nineteenth century,[2] and will probably preserve Henricus Stephanus's name for eternity. He dedicated it to the Emperor Maximilian II, to King Charles IX of France, and to Queen Elizabeth I of England, associating with the names of these sovereigns the universities of their respective countries, with Elizabeth for example the two ancient English universities, Oxford and Cambridge. In spite of all this, the *Thesaurus*, together with the Plato, was his financial ruin, and the last twenty years of his life were spent in a restless hunt for money through the countries and cities of Europe; on one of his travels he died alone in the public hospital of Lyons (1598) and was buried there in French soil.

The achievements of the dynasty of the Étiennes would have been sufficient to secure France the first place in the history of classical scholarship in the second half of the sixteenth century. But there was

---

[1] It has still to be written; meanwhile one has to be grateful for Mark Pattison's brilliant article, 'The Stephenses', *Quarterly Review*, Apr. 1865 = *Essays* I 67–123.

[2] London 1816–28, 9 vols.; cf. the review of G. Hermann, *Opusc.* II (1827) 217 ff.—Paris 1831–65, 8 vols. (vol. I in 2 parts), Didot, by K. B. Haase and the brothers Dindorf.—There have been many discussions about a new Greek *Thesaurus* (see L. Cohn, 'Griechische Lexikographie', Anhang zu K. Brugmann, 'Griechische Grammatik', 4. Aufl., *Handbuch der klassischen Altertumswissenschaft* II 1 (1913) 724 ff.), but we still have to use and be thankful for Henricus Stephanus's work. A new institution at the university of Hamburg undertook in 1944 the task of collecting the complete usage of the Greek language, see B. Snell, *Glotta* 31 (1951) 160. A beginning was made with the *Lexikon des frühgriechischen Epos*, of which 6 fascicles were published between 1955 and 1969, α-ἀπό (1090 col.).

also the family of the Scaligeri and in addition a number of great individual scholars.

When Robert Étienne left Paris for Geneva in 1551, Adrianus Turnebus[1] (1512–65), who had succeeded Toussain[2] in 1547 as Royal Reader in Greek in the Collège de France, became director of the Presse Royale. In this capacity he was able during the next two years (1552–3) to print new texts of Aeschylus and Sophocles; he was far from relying only on his lectures and the oral tradition of his pupils, as Dorat did in these years.[3] In Aeschylus he could use, besides the Byzantine recension of Triclinius, our best manuscript, the Codex Mediceus M (Laur. XXXII 9),[4] which also preserves in its margins much of the ancient grammatical tradition; it had been published earlier in the same year (1552) by Francesco Robortello.[5] Shortly afterwards (1557) another Italian, Petrus Victorius, used a Florentine manuscript (F) to prepare (for the press of Henricus Stephanus) the first complete edition of the *Agamemnon*,[6] of which play two-thirds are missing in the Mediceus and the other manuscripts previously used. There was, as we see, a new interplay between the great scholars of different nations in treating the most difficult Greek texts. There was no longer any need to wait for Greek immigrants;[7] they were superseded and even surpassed by native Western scholars.

Turnebus's edition of Sophocles was based (like his other editions) on critically selected variant readings. He naturally made full use of the manuscripts available to him in Paris, but he did not confine himself to the readings of the family T,[8] as has been supposed. It was his misfortune that Henricus Stephanus, in preparing his Sophocles of 1568, which became the vulgate, both used Turnebus's edition and showed a predilection for T. After his Sophocles Turnebus published in 1553 an anthology of early Greek gnomic poets, Theognis and others. The enormously rich collection of his *Adversaria* (1564/5) gives evidence of his very wide interests, which also included Homer; he published a text of the *Iliad* (1554) and used it for his lectures at the Collège Royal in 1557. His concentration on early Greek poetry was limited to a few

---

[1] See Nolhac (above, p. 102 n. 7) pp. 324 ff. [2] See above, p. 102.
[3] See above, p. 106. [4] Cf. above, p. 48.
[5] G. Toffanin, *La fine dell' umanesimo* (1920); on Robortello and his contemporaries see pp. 29–45 and *passim*.
[6] Aeschylus, *Agamemnon* ed. with a commentary by E. Fraenkel I (1950) pp. 34 f.
[7] See above, p. 102.
[8] Par. gr. 2711, Triclinius' recension; on the problem of the manuscript tradition see H. Lloyd-Jones in his critical review of the text of Sophocles by A. Dain and P. Mazon (1955) in *Gnomon* 28 (1956) 105 ff.; 31 (1959) 478 ff.; 33 (1961) 544 ff.

years of the fifties of the sixteenth century. It is still uncertain whether he was at that time under the spell of Dorat, who as a member of the Pléiade and *poeta regius* composed a highly commendatory poem to introduce his *Adversaria*. On the other hand, Turnebus and Dorat should certainly not be played off against each other; their personalities and the products of their scholarship were different, and each had his proper merits. Turnebus made his conjectures after a more careful consideration of the manuscript tradition; Dorat's emendations were less numerous, but went deeper, as Headlam correctly stated. Dionysius Lambinus, their colleague as Lecteur du Roi, who must be regarded as a competent judge, called them 'paene gemini'. Turnebus's Homeric researches, just mentioned, were included in his *Adversaria*, as were his studies of an important manuscript of Plautus.[1]

After his relatively short 'poetical' period Turnebus turned to Greek philosophy, leading the generation of Ronsard first to Plato and in the course of time also to Aristotle and the Stoics. His comprehensive knowledge of the Greek philosophical sources distinguished his commentary on Cicero *De legibus* and made it famous. He seems to have thought highly of Melanchthon's best pupil Joachim Camerarius;[2] their correspondence is welcome evidence of scholarly intercourse between France and Germany.

But this was still much less active than the intercourse, personal as well as professional, between French and Italian scholars. Dionysius Lambinus (1520–72), who was born in Picardy, lived for about nine years in Italy and used every opportunity of collating manuscripts, ten for his Horace (1561), five for his Lucretius (1564). Lucretius was his favourite poet, and his text with commentary, improved in second and third editions, was highly commended three centuries later by Munro. The whole work was dedicated to King Charles IX, and the several books to his friends, the poets and scholars already mentioned: Ronsard, Dorat, Turnebus, and Muretus. Marc-Antoine de Muret,[3] Lambinus's contemporary (1526–85) and his colleague in the field of Latin scholarship, wrote commentaries on the elegiac poets, on Cicero, Sallust, and on twelve plays of Plautus. Born in France, he had an adventurous life and settled down finally in Rome, where he became a priest. He

---

[1] See F. Ritschl, *Opuscula philologica* II (1868) 4 and 121 on the manuscript and its fortunes. On Camerarius's codices see below, p. 139.

[2] Cf. above, p. 94; see also an autograph Greek letter from Turnebus to Camerarius on the occasion of Melanchthon's death, 23 May 1560, published by Ch.Astruc, *REG* 58 (1945) 219–27.

[3] See above, p. 106. *Opera*, 4 vols. ed. Ruhnken (1789); reprints and selections see Sandys II 152.1.

proved to be an increasingly eloquent and elegant stylist. Montaigne, whose tutor he was, called him 'le meilleur orateur du temps'. Indeed, his writings were regarded as models of good Latin style in all European countries as long as the writing of stylish Latin was considered the highest accomplishment of classical schools. He copied the best ancient prose and could write perfect Ciceronian periods without becoming a servile and narrow imitator. On the other hand, he never achieved the colloquial freedom of Erasmus, which is entirely inimitable. After lecturing for twenty years in Rome, Muretus died there in 1585, in the same year as Victorius—and the final decline of learning in Italy began. But it continued to flourish in France.

Jacques Amyot[1] (1513–93) belonged to the circle of Muretus and Montaigne. In him culminated the long series of illustrious translators of ancient texts into French;[2] his translations of Plutarch (the *Lives* were published in 1559, the *Moralia* in 1572) made him deservedly famous as 'le prince des traducteurs'. This is one of the rare cases in which a translation became an essential part of a modern national literature; Amyot's French Plutarch was eagerly read by Rabelais and Montaigne, and after translation into English (1579), was frequently used by Shakespeare as a dramatic source.

The most important link between the two countries was the family of the Scaligeri from which the foremost scholar of the age descended, Joseph Justus Scaliger (1540–1609). His father Julius Caesar Scaliger (1484–1555), born at Riva on Lake Garda, migrated to France in his forty-second year. In the imposing scholarly works of his son the learned and stylistic achievements of his French and Italian predecessors and contemporaries are combined, and far surpassed. A knowledge of ancient, including Oriental, languages, and of the whole historical material of the ancient world as far as it was preserved in manuscripts, inscriptions, and monuments, was not enough for him; his creative imagination tried to restore lost parts of antiquity. The extraordinary personality of Scaliger was regarded with excessive reverence and excessive hatred; it should always be borne in mind that the opinions of his writings expressed both in his own and in later times were coloured by the acrimony of religious differences. The main part of Scaliger's lifetime coincided exactly with the period of the wars of religion in France (between 1562 and 1598) in which the eve of St. Bartholomew

---

[1] R. Sturel, *Jacques Amyot*, Bibliothèque littéraire de la Renaissance, Sér. I, Tom. 8 (1908); R. Aulotte, *Amyot et Plutarque* (1965) and *Plutarque en France au XVIe siècle*, Études et commentaires 74 (1971).

[2] See above, p. 99 on Seyssel; cf. Sandys II 194 f. 'translators'.

(1572) was only one, though probably the worst, event. The so-called wars of religion were actually a national struggle for power. The French nationalistic spirit, fighting for political unity, and the centralized power of the monarchy, intolerant of any disunity, religious or otherwise, defeated Calvinism politically, and emerged victorious at the end. The literary struggle reflects great discredit upon both sides. Indeed they accused each other of complete lack of integrity. This, however, was going too far. In considering and criticizing the intolerance shown by both sides, we should not deny their integrity in principle; each side firmly believed that it was fighting for truth. It is a sign of Scaliger's greatness that in his scholarly work, though not in his personal writings, he was able to keep out of the political and theological controversy.[1]

Joseph Justus Scaliger believed himself to be descended from a cadet-branch of the princely house of Verona, Della Scala. Although neither exact proof nor refutation of this belief seems to be possible even now,[2] he himself was always confident of his noble ancestry, and felt that his descent imposed an obligation upon him in his life and work. Born in 1540 in the south of France, he went to school at Bordeaux, but only for a very short time; in practice his father Julius Caesar,[3] the author of *Poetices libri septem*, was his principal teacher. He was made to write eighty to a hundred or even two hundred lines of Latin verse every day at his father's dictation and to deliver daily declamation in Latin prose; this practice in speaking and writing gave him a firm grounding in the principles of versification and in the free use of the Latin language. But from his early youth he also had a feeling for the observation of nature, for natural sciences, mathematics, and astronomy, showing himself a true and worthy contemporary of Galileo, Kepler, Tycho de Brahe, and Bacon.[4] His father, while making him a perfect Latin

---

[1] See below, p. 116.

[2] P. O. Kristeller, *American Historical Review* 57 (1952) 394 ff. expressly denied the descent from the Veronese family. On J. C. Scaliger's famous attacks against Erasmus whose greatness was admired by Joseph Justus Scaliger see A. Flitner (above, p. 81 n. 2) p. 97.

[3] *Poetices libri septem* (Lyons 1561), repr. Stuttgart 1964 with an introduction by A. Buck, who tries to fix the position of this most influential theory of poetry in the sixteenth century.— Cf. V. Hall, 'Life of Julius Caesar Scaliger', *Transactions of the American Philosophical Society*, N.S. 40.2 (1950). Short analysis of the *Poetics* with bibliography by Gerh. Jaeger, *Kindler's Literaturlexikon* v (1969) 2229–31 s.v. *Poetices lb. VII*. Cf. also Gerh. Jäger, 'Julius Caesar Scaliger und Joseph Justus Scaliger', *Die Großen der Weltgeschichte V* (1974) 243 ff.

[4] C. M. Bruehl (below, p. 119 n. 7 [on p. 120]) drew attention to the part which the French scientist Guillaume Postel played in Scaliger's life. I have not come across any reference to a relationship between Scaliger and Pierre de la Ramée (1515–72), a classical scholar who based his scientific knowledge on observation (like Scaliger) and contributed to the rise of modern science, see R. Hooykaas, *Humanisme, science et réforme, Pierre de la Ramée* (1958). On scholarship and science see also above, p. 37 n. 8.

scholar, kept him strictly away from Greek language and literature; and it was only at the age of nineteen, after his father's death, that he had the opportunity of going to Paris to learn Greek. At the Collège de France he attended the lectures of Turnebus;[1] but for the most part he remained his own teacher in Greek, reading Homer in three months, all the other Greek poets in the next four months, and in two years the whole of the Greek literature accessible to him.[2] At the same time, in order to practise the knowledge thus acquired, he translated difficult texts like Lycophron and the Orphic hymns (1561) into Latin, making use for this purpose of his astonishing knowledge of early Latin vocabulary. Archaic Latin was his special province; he began at the age of twenty to write his *Coniectanea* (printed 1565) to Varro's *De lingua Latina* and he prepared an edition of Festus (1575), which was first printed years after his death by Scriverius. As a translator he also worked the other way round, from Latin into Greek.[3] Besides passages of Horace, Virgil, the elegiac poets, Martial, and others, he tried his hand at Catullus 66, the *Lock of Berenice*.[4] This was a genuine exercise in Greek verse composition, not an attempt to reconstruct the lost original, Callimachus' Πλόκαμος;[5] Scaliger either did not know or disregarded the first steps that Politian[6] had taken in that direction. In these crowded four years at Paris he began to study Oriental languages, Hebrew and Arabic, which were to be of particular importance for his later work on ancient chronology.

There seems to be no evidence that Scaliger attended Jean Dorat's lectures, but it can hardly be doubted that he was well acquainted with him and with the whole circle of scholars and poets around him. Dorat[7] recommended him to a nobleman of Poitou, Louis Chasteigner de la Roche-Pozay, with whose family he lived, though with considerable interruptions, for thirty years, from 1563 to 1593; and here he found ample leisure for scholarly work. With a member of the Roche-Pozay

[1] There is new evidence about the relation of Turnebus to Scaliger in his Paris years if we can rely on a unique marginal note by Andreas Lucius in a copy of Scaliger's *Epistulae* (1628) in the University of California Library, H. Nibly, *Classical Journal* 37 (1941/2) 293: [Scaliger] 'quem in prima adhuc aetate tantopere admiratus est . . . Hadrianus Turnebus, ut portentosi ingenii iuvenem appellare non dubitaret, ut in epistula quadam ad Meursium scripta Iacobus Gillosus Consiliarius Gallicus testatur (instatur MS.).'

[2] Cf. Winckelmann's voracious private reading below, p. 168.

[3] J. J. Scaliger, *Poemata omnia*, 2nd ed. 1864, pp. 163–257 'Graece reddita', first printed seven years after his death by Scriverius, see Mark Pattison, *Essays* I p. 216.

[4] *Poemata omnia* (above n. 3) pp. 214 ff., dedicated to Muretus, Paris 1562.

[5] Call. fr. 110 with Addenda I and II.

[6] See above, p. 45, and Call. II p. xliii.

[7] Pattison, *Essays* I 218 ff.; on Dorat and Scaliger see also Nolhac (above, p. 102 n. 7 [on p. 103]) p. 202.

family he travelled through Italy, and in Rome he met Muretus,[1] for whose help he remained grateful for the rest of his life. Scaliger was especially interested in inscriptions and collected as much material as possible, which he made available to Janus Gruter in later years for his *Corpus inscriptionum antiquarum*, published in 1602; in this book not only a great many of the inscriptions, but also the twenty-four methodical indexes are Scaliger's work.

He disliked the Italians, who were, as he thought, frivolous atheists, to whom the classics were only playthings; he felt the deepest aversion to papal Rome. Though brought up in the Catholic faith, Scaliger had come into close contact with Calvinistic circles in his Paris years, and he apostasized either in 1562 before his Italian journey or afterwards in 1566. It is understandable that he should have detested the growing political struggle being waged under the pretext of religion and that some Calvinistic ideas should have appealed to him. He believed that he found in them a spiritual independence, an impetus to real criticism as an instrument of truth; but since 'he did not dispute on the controversial points of faith', as his greatest friend the Catholic historian de Thou said, it is almost impossible to come to a conclusion about his beliefs. One thing, however, is certain: he had a profoundly religious mind and embraced 'Muse and religion' with equal love. On the relation of *grammatica* and *religio* there is a very remarkable dictum in the *Scaligerana*,[2] of which only the first part is usually quoted,[3] perhaps his most famous words: 'Utinam essem bonus grammaticus.' But the meaning is unmistakably given by the passage that follows: 'Non aliunde discordiae in religione pendent quam ab ignoratione grammaticae', all controversies in religion arise from ignorance of *grammatica*. This is not 'grammar' in the trivial sense, but criticism in the Hellenistic sense of γραμματική as the κριτικὴ τέχνη.[4] When we look back to Erasmus and his contemporaries and pupils, we can hardly deny that Scaliger touched one of the chief problems of his century. But he did not apply his scholarship *in extenso* to this problem himself.[5]

After his Italian journey he went to England, where he made one of his biting remarks on the indolent lives of fellows of colleges in Cambridge in connection with his note on a Cambridge manuscript of Origen's *Contra Celsum*; he was disappointed in his search for Greek

---

[1] See above, p. 106 and p. 115 n. 4.

[2] *Scaligerana* ed. alphab. Colon. (1595) pp. 176 f.; cf. Bernays, *Scaliger* (below, p. 119 n. 7) p. 19.

[3] 'Halfquotations', see above, p. 73 n. 8.

[4] See *History* [1] 299 s.v. grammar and 308 s.v. γραμματική.   [5] Cf. above, p. 114.

manuscripts, but he was at least able to borrow a copy of the *Lexicon* of Photius from a Cambridge friend. Returning to France during the height of the civil war, he lived an unsettled camp life with the family de la Roche-Pozay. In 1570 he went to Valence in the Dauphiné where Jacques Cuiacius (1522–90), the greatest expert in Roman law,[1] was teaching at that time. They became intimate friends, and Scaliger acquired a considerable knowledge of civil law, for him a new instrument of historical and philological inquiry. It was at Valence that his lifelong friendship with de Thou began. In 1572 he left Valence before the fatal eve of St. Bartholomew and remained for two years in Switzerland, where he was offered a chair of philosophy in Geneva. But he had no taste for lecturing and a special aversion to philosophy; it is noticeable how relatively few quotations from Plato there are in his writings. As a Huguenot, he should have felt at home in Calvinistic Geneva, but he was soon alienated by the narrow sectarian spirit and the tyrannical policy of the pastors. So after two years he returned to the disorders and frequent dangers of the civil wars in France. There, as it turned out, for the next twenty years, staying with his friends in French castles, he enjoyed a relatively quiet life for his scholarly work. Finally the university of Leyden called him to Holland, and after negotiations lasting three years he went there in 1593, the year in which Henri IV returned to the Catholic faith. In Leyden only his presence was required without any obligation to teach and before his peaceful death in 1609 he was able to finish his monumental works.

Nobody before Scaliger had any proper knowledge of the *early* Latin language;[2] and when he turned to the *Appendix Virgiliana* (1572) and to the three elegists, Catullus, Tibullus, and Propertius (1577), he applied his unique knowledge of the archaic language to these later poems. His editions contain ingenious emendations as well as violent mistakes. His Manilius of 1579 marked the transition into his proper field of research, that of historical reconstruction: his real object in the Manilius was not to restore the text of that extremely difficult astronomical epic by emendatory criticism and skill in language, but to reconstruct the astronomical system of the first century A.D. This led on naturally to his next work, the *De emendatione temporum*, first published 1583 in a folio volume, much improved in the second edition of 1598, and finally enlarged to the *Thesaurus temporum* in 1606. He used the improved astronomy of his age as a scientific basis for historical

---

[1] On jurisprudence cf. above, pp. 86 and 101.
[2] See above, p. 115.

chronology.[1] His view—which anticipated much of F. A. Wolf's conception of 'Altertumswissenschaft'—was that the history of the ancient world had to be known as a whole, if at all. With his knowledge of all ancient languages, classical and Oriental, as well as of ancient history, Oriental, Greek, Roman, and Christian, he undertook to reconstruct all the chronological systems of the ancient world. For this purpose he had to collect the fragments of the chronologists of late antiquity; the groundwork for the study of the tradition was Jerome's translation of the chronicle written by Eusebius at the time of Constantine. All the Renaissance editors of St. Jerome, even Erasmus in his nine-volume edition,[2] had omitted the chronicle, because it was completely unintelligible to them. Scaliger conjectured that Eusebius' work originally had two books, of which Jerome had translated only the second with the chronological tables, and that the original of the first could perhaps be restored with the help of the Byzantine historian Georgios Synkellos.[3] A Paris manuscript of Synkellos was finally sent to Leyden in 1602, and with incomparable boldness, Scaliger reproduced the whole original work of Eusebius, restoring the first book from the Byzantine excerpts and retranslating the second from Jerome whose many mistakes he took the opportunity of correcting. In 1605, shortly before the final edition of the *Thesaurus temporum*, Casaubon found in the Paris library a chronological list of Olympian victors; Scaliger recognized it as the compilation of Julius Africanus, used by Eusebius in his first book. Surprising discoveries thus helped to verify ingenious combinations. But the greatest surprise came two centuries afterwards, one of those very rare occasions in the history of scholarship when a later find produces the evidence which confirms the principle of a hypothetical historical reconstruction. A fifth-century Armenian version[4] of Eusebius' chronicle, published in 1818, revealed that Scaliger's intuition had guided him aright in plotting his amazing chronological work.

But in the first half of the seventeenth century vehement polemics started, which culminated in the *Opus de doctrina temporum* of the learned Jesuit Denys Petau (3 volumes, 1627–30). The astronomical knowledge of this formidable opponent was still more solid than that of Scaliger; he was also more cautious in his conjectures and was able to correct a number of Scaliger's mistakes.

[1] Cf. *History* [1] 163.   [2] Cf. above, p. 78.

[3] The German rediscoverers of Byzantine literature, from Hieronymus Wolf to Höschel, actually followed Scaliger's lead.

[4] Preserved in two manuscripts and edited by J. B. Aucher, 2 vols., Venice 1818.

Neither Italy nor France showed real understanding of Scaliger's genius or much enthusiasm for its creations. But in Leyden a circle of devoted admirers and pupils surrounded him. In the field of early Latin literature Janus Dousa (1545–1604), the first curator of the new university, turned to Plautus (1587), and his younger son Franciscus was inspired to collect the fragments of Lucilius (1597). Daniel Heinsius (1580–1655) became Scaliger's favourite Dutch pupil; and Scaliger was the first to recognize the rare gifts of the young Hugo de Groot, the greatest Dutch scholar of the seventeenth century. Looking further afield we find his chronological labours immediately appreciated in England by John Selden, the learned editor of the so-called *Marmor Parium*, two fragments of a chronological table found at Paros and now in the Ashmolean.[1] We have already noticed his connection with some contemporary scholars in Germany.[2] And it was there—but not before the middle of the eighteenth century—that Winckelmann[3] recognized in Scaliger the first scholar to have restored the life of the ancient world in its totality; Winckelmann read not only the ancient texts, but also the great French classical scholars, Scaliger himself, his friend de Thou, and his pupil Grotius, as we can tell from his references to them. In his *History of Ancient Art*,[4] first published in 1764, he took up the suggestion made by Scaliger in a letter of the year 1607 to Salmasius[5] that there were in his opinion four ages of Greek poetry, and accepted the same sequence of four distinct ages for the evolution of Greek art. The German concept of a general science of antiquity[6] in the following generations is based on Winckelmann's ideas, and in this respect his own reference back to Scaliger is of signal historical importance.

After this digression to look at the appreciation and the effect of Scaliger's works[7] in other countries we must return to follow the

---

[1] *Marmora Arundeliana* (1628/9).

[2] See above, p. 118 n. 2.

[3] I have been unable to find a direct reference to this in the literature on Scaliger. C. Justi, *Winckelmann und seine Zeitgenossen* I (1898) made occasional references, esp. pp. 134 and 160, but not to the passages quoted below, nn. 4 and 5.

[4] 'Geschichte der Kunst des Altertums' 8. Buch, 1. Kap. = *Werke* I (Suttgart 1847) 299.

[5] *Epistolae* (Francofurti 1628) pp. 486 f.

[6] I expected to find something on Scaliger in F. A. Wolf's writings, but have not so far succeeded. The enthusiastic utterances of Niebuhr and his contemporaries are quoted again and again.

[7] There is no collection of *Opera omnia*; we have to refer to the titles of the individual books.—*Epistolae omnes*, ed. D. Heinsius, Lugd. Bat. 1627, reprinted Frankfurt 1628 (the edition I use). *Lettres françaises inédites* publ. et annotées par Ph. Tamizey de Larroque (1879). —*Scaligerana ou bons mots* I Vertuniani (1574–93) and II Vassanorum (1603–6) (names of the compilers), best publication 2 vols. by Pierre des Maizeaux, Amsterdam 1740; a new critical edition was prepared by M. Bonnet, and then by M. A. Monod (1920?), but nothing has

development of scholarship in France after Scaliger. His particular greatness becomes still more evident when we turn to his younger contemporary Isaac Casaubon (1559–1614) and to his successor in Leyden, Claudius Salmasius (1588–1653). Casaubon was born in Geneva, the son of a Huguenot pastor, and had a hard youth, spent latterly in the mountains of the Dauphiné with his father, who was his only teacher. At the age of twenty he came back to Geneva to pursue his Greek studies more intensively; he also married one of Henri Étienne's many daughters and had nineteen children. After lecturing in Geneva and Montpellier, he was invited to Paris in 1599 by King Henri IV, who desired to restore religious peace. Here he was given the title 'Lecteur du Roi', but held no official position either in the university or in the Collège de France until he became sublibrarian to De Thou in the Royal Library. When Henri IV was murdered in 1610, Casaubon was urged to become a Catholic; though he was far from sympathizing with the narrow dogmatism of Calvinism in his Paris years, he could not make up his mind to join the Roman Church. He was in favour of a *via media* and accepted with pleasure the invitation of the archbishop of Canterbury to England. Shortly afterwards, in 1614, he died, exhausted by his excessive learned labours at the age of fifty-five, and was buried in Westminster Abbey.[1]

been published.—At first sight it seems rather strange that the only monograph is still Jacob Bernays, *Joseph Justus Scaliger* (Berlin 1855), but only pp. 1–17 and 31–104 contain the actual introduction and life, by far the greater part being filled by the notes and the appendices; but considering Scaliger's greatness one understands why no one should have felt himself equal to undertaking a fully documented biography. Bernays dedicated his work, more an essay than a book, to F. Ritschl, who regarded him as his most gifted pupil. Cf. Wolfgang Schmid, 'Friedrich Ritschl und Jacob Bernays', *Bonner Gelehrte*, Beiträge der Geschichte der Wissenschaften in Bonn, Philosophie und Altertumswissenschaften (1968) p. 137, and especially A. Momigliano, 'Jacob Bernays', *Mededeelingen der K. Nederlandse Akademie van Wetenschappen, Afd. Letterkunde* N.R. 32. 5 (1969) 151–78. A great scholar himself, Bernays was able to recognize in Scaliger's work the absolute perfection of knowledge and method, and we must be grateful for his sensible and noble *Scaliger*. It was reviewed by the only competent critic Mark Pattison (to whose publications I referred, *History* [I] p. ix) in *Quarterly Review* 1860 (reprinted in *Essays* I [1889] 132–95). In consequence of this review it was suggested to Pattison by Bernays' great friend Chr. K. J. von Bunsen in a conversation during 1856 that he should 'write Scaliger's life in connection with the religious history of the time'. For nearly thirty years Pattison tried to get together the material (*Memoirs*, 1885, pp. 321 ff.); having looked through his notes in the Bodleian Library (cf. above, p. 102 n. 7 [on p. 103]) I regret to say that he would never have been able to execute this ambitious plan. Nor could anyone else; see the survey of efforts and reasons for failure given by C. M. Bruehl, 'J. J. Scaliger. Ein Beitrag zur geistesgeschichtlichen Bedeutung der Alterumswissenschaft', *Zeitschrift für Religions- und Geistesgeschichte* 12 (1960) 202–18 and 13 (1961) 45–65. J. Scaliger, *Autobiography* transl. by G. W. Robinson (Cambridge 1927) contains useful excerpts from his letters in English translation with introduction and notes.

[1] Mark Pattison, *Isaac Casaubonus* (1875); a second edition of this book of more than 500 pp. was issued in 1892. It is a detailed, well-documented, sympathetic modern biography such as

Casaubon was a scholar of a highly individual kind;[1] not being an outstanding grammarian and critic, he did not become in the first place an editor of critical texts, and not having an inventive imagination, he made no historical reconstructions. He was a patient reader and collector; and his genius, if the word is allowable in this connection, was for untiring mental effort. His aim was to amass exhaustive knowledge through extensive reading of all possible sources, and then to construct a picture of the ancient world by putting together what he had learnt. He was always in a state of despondency, because he was for ever finding new texts and new books and was afraid that time would not allow him to perfect his knowledge. His mission was to write commentaries,[2] of which the most important were those on the *Geographica* of Strabo (1587, second edition 1620),[3] on Theophrastus' *Characters* (1592, second edition 1599, third 1612, and many thereafter), on Suetonius (1595),[4] and on Athenaeus (1600).[5] His *Animadversiones* on the last of these were written with groaning and sighing, day and night, through more than three years. Nobody since Casaubon has possessed self-denial enough for making commentaries on texts like Strabo or even Athenaeus. His notes on Polybius' Greek text, with a Latin translation, were published posthumously in 1617; a new point he made in the preface was that history like that of Polybius is the preparatory school for the politician; the influence of Dutch[6] scholarship on Casaubon's last work can hardly be doubted.

His commentary of 1605 on Persius' *Satirae* (still reprinted verbatim in 1833) was accompanied by an essay 'De satyrica Graecorum poesi et Romanorum satira' (often reprinted separately until the end of the eighteenth century); this is the first monograph on a problem of the history of ancient literature. All his writings were based on material independently collected by himself; that distinguishes them from the

hardly exists of any classical scholar, except perhaps Walser's Poggio (see above, p. 31 n. 2); but it does not quite do justice to Casaubon's scholarly achievements (a weakness shown by Walser's monograph on Poggio). His letters and diary have been printed, *Epistolae* (3rd ed. 1709) and *Ephemerides*, 2 vols. (Oxford 1850). Manuscripts: 60 vols. of 'Adversaria', are preserved in the Bodleian Library; some of his books, especially texts with marginal notes, are in the Cambridge University Library, for instance the Aeschylus with Scaliger's as well as his notes, but his Polybius is in the Bodleian.

[1] He was still under the influence of Dorat, and his manuscript notes on Greek tragedy show that it is unjust to call him an ἄμουσος, see E. Fraenkel, Aesch. *Ag.* 1 38 and 77.

[2] A complete list in Mark Pattison (above p. 120 n. 1) 534 ff.

[3] We still quote the pages of this edition.

[4] See S. Weinstock, *Divus Julius* (Oxford 1971). Casaubon's merits as textual critic are shown by Weinstock.

[5] Re-edited by Schweighäuser 1801 and reprinted eight times till 1840.

[6] See below, p. 126.

mere compilations of the later seventeenth and eighteenth centuries, especially in Holland and Germany, and there has been no one else in the history of classical scholarship whose commentaries remained indispensable and unsuperseded for such an exceedingly long time. In textual criticism he was rather conservative, insisting first of all on the authority of manuscripts, but not afraid of bold conjectures. The most striking fact, however, for anyone coming to him from the Italian and northern humanists or from the French circle of the Pleiad will ever be this: that Casaubon is the first pure type of a classical scholar destitute of sympathy for human and aesthetic values. He is the best and perhaps the greatest example of the ascetic brand of scholar who sacrifices his life for his high purpose.

Claudius Salmasius[1] (1588–1653), born at Saumur, spent his early youth in Paris under Scaliger and Casaubon, who described him at nineteen as a 'iuvenis ad miraculum doctus'.[2] He certainly deserved his reputation for learning as the Eratosthenes of his time;[3] his work was marked by an inclination to the abstruse, and his enormous reading was not always in the service of the explanation of the text, but sometimes amounted merely to the accumulation of material for its own sake.

He wrote on a bewildering variety of subjects: for instance *De usuris* (1638), a treatise which after a historical survey insists on the legitimacy of usury for clergy and laity, *De caesarie virorum et mulierum coma* (1644), *De primatu papae* (1645), *Defensio Regia pro Carolo I* (1649).[4] He is generally supposed to have 'discovered' in Heidelberg (whither he moved from Paris in 1607) the codex Palatinus of the *Greek Anthology*, but it may only be that Janus Gruter[5] showed him the codex and that Scaliger urged him constantly but in vain to publish it. What he did work at in his years at Heidelberg was his *Plinianae exercitationes* (1629), perhaps his most remarkable book, which deals mainly with Solinus' excerpts; there he established by acute observation a rule about the components of local names like $\Pi\epsilon\rho\sigma\epsilon\pi o\lambda\iota\varsigma$ which proved to be absolutely correct.[6] In 1632, after the chair had been vacant for twenty-three years, he became Scaliger's successor at Leyden. There he found the leisure to

---

[1] Gustave Cohen, *Écrivains français en Hollande dans la première moitié du XVIIe siècle* (Paris 1920) pp. 311–34 'Le plus grand philologue du XVIIe siècle: Claude Saumaise'. But there is nothing in this chapter about the 'Anthologia inedita' or the *Lingua hellenistica*.

[2] In a letter to Scaliger (*Epp.* p. 284).   [3] Cf. *History* [1] 170.

[4] Written at the request of the exiled Charles II, this was the starting-point of a controversy with Milton, whose reply *Pro populo Anglicano Defensio* (1651) provoked in its turn a rejoinder by Salmasius.

[5] On Gruter and Scaliger see below, p. 138.

[6] J. Wackernagel, *Glotta* 14 (1925) 36 ff. = *Kleine Schriften* II (1953) 844 ff.

publish the series of books already mentioned, to which must be added the treatise *De lingua Hellenistica* (1643);[1] his justified arguments against the assumption, maintained by several scholars in Scaliger's day, that the Greek of the New Testament was a special dialect had the paradoxical effect that the name 'lingua Hellenistica' became more popular and could still be found in Greek grammars of the early nineteenth century. Indeed the use in Buttmann's *Ausführliche griechische Grammatik* I (1819) 7 n. 12 suggested to Droysen the name 'Hellenistic age' to describe the centuries between Alexander and Augustus.[2]

In 1650 Salmasius left Leyden for the court of the Swedish Queen Christina,[3] the daughter of Gustavus Adolphus, where he died shortly afterwards. The influx of French writers in the first half of the seventeenth century was a blessing for Holland. The greatest of them was Descartes, and the contingent of classical scholars was a relatively small, but it was an important one. It is therefore natural to turn now from France to Holland and to come back to France later.

[1] The book *Funus linguae Hellenisticae*, often ascribed to Salmasius, is the work of an anonymous writer.

[2] See *Ausgewählte Schriften*, pp. 150 f.

[3] *Christina Queen of Sweden. A Personality of European Civilization* (Exhibition, Nationalmusei Utställingskatalog 305, Stockholm 1966, 622 pp., 96 plates) pp. 204 ff. Foreign scholars at the court.

X

CLASSICAL SCHOLARSHIP IN HOLLAND AND IN POST-RENAISSANCE FRANCE, ITALY, AND GERMANY

THE northern provinces of the Netherlands, after their long war for freedom, declared their independence in 1579; and they did so in the university of Utrecht. At Leyden after the expulsion of the Spanish troops in 1575 a new university was founded which soon began to attract many great scholars.[1] These events initiated the golden age of Holland in art and literature and learning during the seventeenth century.

Its cultural life was at first in closer contact with France and the southern provinces than with Germany, and Calvinism, not Lutheranism, predominated. But it is characteristic that one of the Leyden professors, Jacobus Arminius (1560–1609, at Leyden from 1605 onwards), rejected Calvin's dogma of predestination.[2] For, although Arminius's own doctrine was then condemned by a Calvinistic synod, and his followers, the Arminians, were subjected to persecution, there seems to have been no active Calvinist amongst the greatest figures of Holland's prime in the seventeenth century.[3] The religious situation, in fact, was complicated, not to say confused, and an extreme example of this was the first classical scholar[4] invited to Leyden in 1579, Justus

[1] Cf. above, pp. 117, 119, 122 f.; on classical scholarship in the Netherlands in the fourteenth and fifteenth centuries see above, especially p. 70 (R. Agricola and Erasmus) and p. 106 (W. Canter and Jean Dorat).—Cf. H. Schneppen, *Niederländische Universitäten und deutsches Geistesleben* (1960) pp. 116 ff. Leiden, on the scholars (sixteenth to eighteenth century) see Index. A. Gerlo and H. D. L. Vervliet, *Bibliographie de l'humanisme des anciens Pays-Bas* (1972). *The Leiden University in the Seventeenth Century* (Leiden 1975) pp. 161 ff: J. H. Waszink, 'Classical philology'.

[2] On Arminius and his relations to classical scholars, to Scaliger and others, see C. M. Bruehl (above, p. 119 n. 7 [on p. 120]), XIII (1961) 48 ff.

[3] J. Huizinga, *Holländische Kultur im 17. Jahrhundert* (1933) p. 32; cf. G. N. Clark, *The Seventeenth Century* (1929, 2nd ed. 1947), on Calvinism pp. 310 ff.—A lively survey of the complex character of the seventeenth century, and characteristic details also of scholarship and science, are given by E. Gothein, *Schriften zur Kulturgeschichte der Renaissance, Reformation und Gegenreformation*, Bd. 2: *Reformation und Gegenreformation* (1924), with selected notes.

[4] Lucian Müller, *Geschichte der klassischen Philologie in den Niederlanden* (1869, repr. 1970) still gives quite a useful survey in its first part, 'Die philologischen Schulen der Niederländer', pp. 1–129; but it needs to be supplemented by Waszink's excellent paper (above, n. 1).

Lipsius (1547–1606).[1] Born in the Spanish Netherlands, he studied Roman law at Louvain,[2] the Catholic university founded in 1517 with its famous *collegium trilingue*, and becoming secretary to Cardinal Granvella he accompanied him on his return to Italy. Later he visited the imperial court at Vienna, but then settled down in the strictly Protestant university of Jena and became a Lutheran himself. While remaining a Lutheran he married a Catholic and returned to his Catholic native land, lecturing for some time at Louvain. But he did not hesitate to accept the invitation to the new Dutch university of Leyden which was officially Calvinistic; he remained there for twelve years as honorary professor of Roman history and antiquities; then he took leave, went to Mainz, returned to the Catholic faith, and for the last fourteen years of his life lectured on history in his first university, Louvain. If this sketch suggests a caricature of religious, or at least denominational, indifference on Lipsius's part, it also shows a welcome tolerance in the various seats of learning with whose support he became the first of living Latin scholars and critics.

It was characteristic of Holland, in contrast to France, that there was much less interest in Greek than in Latin.[3] Willem Canter (1542–75),[4] having been under the strong influence of Dorat in Paris, was an exception. In his successive editions of the tragedians, Euripides (1571), Sophocles (1579), Aeschylus (1580)—the last two published after his early death—he outdid Turnebus[5] by distinguishing and marking the metrical responsions in the choruses. Otherwise, the study of Latin prevailed, but it was confined to a special field. The favourite authors were not Cicero, not Livy, not Virgil, not Horace, but Seneca, Tacitus, Lucan, and even Claudian; and interest was concentrated less on their language and style[6] than on their political attitude. Already Salutati[7] had begun to interpret the classical writers politically, while Politian[8] had turned to post-classical authors, and Beatus Rhenanus[9] to the later Roman historians. The first political commentary on Tacitus seems to

---

[1] *Opera omnia*, 4 vols. (Antwerp 1637). V. der Haeghen, *Bibliographie Lipsienne*, 3 vols. (Gent 1886–8). *Inventaire de la correspondance de J.L.* par A. Gerlo et H. D. L. Vervliet (Antwerp 1968).

[2] Cf. above, p. 96.

[3] The influence of his studies of Roman antiquities seems to have reached Karl Otfried Müller who turned them into genuine historical researches, see W. Kaegi, *Deutsche Zeitschrift* (Jg. 49 *Kunstwart*) 1935/6, p. 97.

[4] See above, p. 106; cf. Nolhac (above, p. 102, n. 7), p. 212 and notes 2–4.

[5] See above, p. 111.

[6] But see M. W. Croll, *Studies in Philology* 18 (1921) 79 ff.

[7] See above, p. 25.

[8] See above, p. 44.   [9] See above, p. 84, on Tacitus especially.

have been written by Carolus Paschalius in 1581,[1] and this started a 'Tacitean movement' which spread from Italy across Europe and reached its height in the work of Lipsius.[2] His first edition with commentary of 1574 was much enlarged in that of 1600, for which he was able to use Paschalius and Muretus, and it was followed by two posthumous editions. He also edited other late Roman historians, Valerius Maximus and Velleius Paterculus.

With his Seneca (1605) Lipsius tried to promote the knowledge of Stoic philosophy. Stoic maxims had been occasionally recommended by humanists of the fifteenth and sixteenth centuries, but Stoicism had not been able to compete with the Christian Platonism[3] entrenched in Italy and in transalpine countries. Now, however, the ideal of Christian Stoicism was proclaimed by Lipsius and others in special treatises and had considerable influence upon the philosophy of the age.[4] Their object was to combine all their knowledge of Roman history and their skill in the heroic eloquence of the later Roman authors with Stoic philosophy in the foundation of a new 'doctrina civilis', aiming not at independent and self-sufficient scholarship or 'humanitas', but at the education of the 'homo politicus'. We cannot understand the classical studies of Lipsius and his contemporaries, if we overlook the importance of this new political intention. The fundamental document of this approach is Lipsius's *Politicorum sive civilis doctrinae libri sex* (1589, enlarged in two later editions, 1596 and 1605).

The greatest and most catholic-minded *homo politicus* in the series of Dutch scholars was Hugo Grotius (1583–1645).[5] Born at Delft, he became a pupil at Leyden of Scaliger, who recognized the genius of the infant prodigy; he could write Latin verse by the time he was eight (in the course of his life he was to produce more than 10,000 verses in Latin[6] and about the same number in Dutch). At fifteen he finished his commentary on Martianus Capella; at seventeen he published the

---

[1] A. Momigliano, 'The First Political Commentary on Tacitus', *Contributo* [I] (1955) 36–59, esp. 40 ff.

[2] C. O. Brink, 'Justus Lipsius and the Text of Tacitus', *JRS* 41 (1951) 32–51 gives a fair appraisal of Lipsius's achievement and those of his predecessors.

[3] See above, pp. 57 ff.

[4] See W. Dilthey, *Ges. Schriften* II 443 ff.; cf. J. L. Saunders, *J. Lipsius, The Philosophy of the Renaissance Stoicism* (New York 1955) who gives a full analysis of Lipsius's theory and of its influence.

[5] See Sandys II 315–19; I am indebted to this exceptionally good section; cf. B. A. Müller, *GGA* 186 (1924) 18 ff. in his extensive critical review of R. Helm, *Hugo Grotius* (Rektoratsrede Rostock 1920).

[6] O. Kluge, *Die Dichtung des Hugo Grotius im Rahmen der neulateinischen Kunstpoesie* (1940).— H. Grotius, *Briefwisseling* I (1928)—IX (1973), to be continued.

Greek text of Aratus' *Phaenomena* with the ancient Latin translations ingeniously supplementing the corrupt fragment of Cicero's translation. Later in his life he translated into Latin the epigrams of the *Anthologia Planudea*,[1] hundreds of excerpts from Greek dramatists, and the *Phoenissae* of Euripides,[2] edited Lucan (1614, with 'notae') and Silius Italicus (1636), and published conjectures to Seneca and Tacitus. But that was only a small part of his life's work. After taking his degree in law[3] (*De iure praedae*, 1604/5), he started his public career as the official historiographer of the Netherlands, writing the *Annales* and the *Historiae* as the Tacitus of his country; then he became advocate general of Holland and Zealand, a member of the States-general, and envoy to England. At the age of twenty-two he had begun to work on international law, the subject of his most famous publications, of which *Mare liberum* (1609) was the first (to be answered by Selden in the *Mare clausum*, 1636) when his career was interrupted. In the theological field he openly sympathized with the moderate group of Arminians.[4] When their opinions were condemned by the synod of Dordrecht in 1619, his friend Barneveldt was sentenced to death, and Grotius himself was condemned to imprisonment for life (in contrast with the tolerance shown towards Lipsius). In prison he wrote his most popular Dutch poem 'The proof of true religion for the use of the Dutch sailor'. His own translation into Latin, published later in Paris, 'De veritate religionis Christianae', spread through the whole world and was translated into many modern languages; Leibniz called this poem a 'livre d'or' in which Grotius had surpassed himself and all his contemporaries. After a year and ten months in prison Grotius escaped in a large bookcase and fled to Paris. There in 1625, the seventh year of the Thirty Years' War, he wrote the book with which his name is connected for ever in the modern world, *De iure belli ac pacis*. His philological, juridical, historical, and theological knowledge is combined in this classic work, which is not only the first concept of international law, but also the work of a classical scholar in the Erasmian spirit. Erasmus had tried to preserve the endangered catholicity of religion and learning, and had devoted all his scholarship and humanism to the promotion of the universal church and its spiritual leadership of the whole of Christendom. But unity had

---

[1] Reprinted in the Didot edition of the *Anthology*.

[2] We must be conscious of the circumstances in which Grotius composed these bulky translations as a prisoner.

[3] On classical scholarship as the basis of the study of jurisprudence see above, p. 86 and notes with further references.

[4] See above, p. 124.

been lost for a century when Grotius wrote *De iure belli ac pacis*. Though this is the work of a truly religious man, it is nevertheless a secularization of the Erasmian ideas that he presents as he tries to construct in the place of the lost universality of the church a new unity, a Christian humanistic society of nations. Belief in the eternal human values and the wisdom of the ancients, and belief in the divine truth of the Gospels are the foundations of this structure of international law; Grotius fervently hoped that peace and reunion would arise in the end and that only a 'good interpreter'[1] would be necessary. We may understand that such views were subject to suspicion by the Protestant world in which he lived. His defence in his own case, published in Dutch and in Latin, and his attempts to return from France to his native land were met by a decree of perpetual banishment. It may seem to us paradoxical that King Gustavus Adolphus of Sweden should have become an admirer of Grotius's book on international law. But, in consequence, Grotius entered the service of Sweden and became for many years the envoy of Queen Christina[2] at the court of France.[3] On a journey back from Stockholm to Paris his ship was wrecked in the Baltic Sea, and he died at Rostock in 1645. Joost van den Vondel in one of his most moving poems celebrated the personality of the great scholar, his dearest friend. In a letter about the *pax Christiana* Grotius had said that, even if he were not to be allowed to enjoy the fruit of his labours, it was his duty none the less 'serere arbores alteri fortasse saeculo profuturas', 'to sow trees which may be possibly useful for another century'. We still hope and wait for that century.

When we look at the considerable number of good Dutch scholars contemporary with Grotius or belonging to the following two generations we notice a fact[4] unique in our history: that classical scholarship was hereditary in certain families. We have already mentioned Janus and Franciscus Dousa;[5] they were followed by two Heinsii, two Vossii, two or even three Gronovii (who came over from Hamburg), and in the following century by two Burmanni, uncle and nephew. We can do no more here than say a little about the main tradition.[6]

[1] An Erasmian formula often repeated.   [2] See above, p. 123 n. 3.
[3] Richelieu respected, but did not like him.
[4] The case of the few outstanding members of the Stephani and the Scaligeri is different.
[5] Above, p. 119.
[6] Attention should be drawn to a book in which nobody would expect to find a review of the multifarious critical literature of the seventeenth century: Stanislaus von Dunin-Borkowski, *Spinoza*, vol. IV (1936) 'Aus den Tagen Spinozas, 3. Buch: Das Lebenswerk'. This is a proof of the author's remarkable learning that he gives this 200-page review of classical and biblical criticism as the background for Spinoza's *Tractatus theologico-politicus* (1670). Unfortunately he

The treatment of Latin literature in the tradition of Lipsius was continued by Daniel Heinsius (1580–1655), Casaubon's 'parvus Scaliger',[1] and by his son Nicolaus (1620–81). Daniel and Nicolaus were Latin poets and preferred the study of Latin literature;[2] but Daniel went from Horace's *Ars poetica* back to Aristotle, and his *De tragoediae constitutione* became in its time one of the most influential books on the *Poetics*, which he also edited and translated (1611). His criticism of Latin poets, however, above all of Ovid (1629), has lasting value. His son and pupil Nicolaus was a keen traveller and collated many manuscripts abroad when he was in diplomatic service; he showed a genuine sense of poetry in the selection of the variant readings for his numerous editions of Latin poets from Virgil to Claudian which became the foundation of all the later critical texts. In this respect he was not unjustly praised as 'sospitator poetarum Latinorum'.

In this group of textual critics and editors we may include the elder Gronovius, Johann Friedrich (1611–71); working at Leyden from 1634 onwards as a pupil of Daniel and a friend of Nicolaus Heinsius, he devoted himself mainly to later Latin prose. So did the younger Isaac Vossius (1618–89). But the elder Gerard John Vossius (1577–1649) and the younger Jacob Gronovius (1645–1716) belong to a quite distinct group of antiquarians[3] and 'polyhistors'.[4] Their contribution to classical scholarship consisted in enlargement and consolidation rather than originality. But to put in order and make easily accessible the accumulated treasures of the creative age from Agricola and Erasmus to Grotius was no mean service,[5] and later generations have been duly grateful for it. When G. J. Vossius published a work proposing a system of polymatheia (1650) he had some justification for calling it *De philologia*.[6]

At the beginning of the seventeenth century the intolerant fanaticism that had led to the horrors of the religious wars was confronted by a revival of Christian humanism, that 'devout humanism' of which

died before the volume was printed, and it may not be his fault that this chapter is a little untidy; but it deals conscientiously with books characteristic of the age which hardly anybody has read or even opened for centuries.

[1] Cf. above, p. 119.

[2] Cf. above, p. 125. See H. J. de Jonge, *Daniel Heinsius and the Textus Receptus of the New Testament* (Leiden 1971).

[3] See A. Momigliano, 'Ancient History and the Antiquarian', *Contributo* I (1955) 67 ff.

[4] On πολυΐστωρ and related words see *History* [I] 125.5.

[5] Huizinga, *Holländische Kultur* (above, p. 124 n. 3) p. 61, overlooks this point when he speaks only of the stiffening of Dutch civilization towards the end of the seventeenth century.

[6] Cf. what was said above, p. 101 on Budé's *De philologia* and *History* [I] 158 f. and 170 on Eratosthenes.

St. François de Sales became the greatest representative. The earlier Christian humanism, despite its popular tendencies, had been more for the educated classes; but devout humanisme tried to bring its principles and its spirit within the reach of all, stressing the holy and its practice rather than beauty and truth. So it did not directly inspire scholarly work, but prepared for it a new atmosphere of peace and reason and a balance of divine and human values. In this new spiritual atmosphere the learned work of the monastic orders began to prosper again, and towards the end of the century the oratory and the literary work of great preachers and writers. The importance of devout humanisme was stressed or probably discovered by Henri Bremond.[1]

Strict scholarly criticism in the ecclesiastical field was promoted by the biblical studies of the French Oratorian Richard Simon (1638–1712), 'Father of biblical criticism'.[2] From his *Critical History of the Old Testament* (first edition 1678, English translation 1682) he drew revolutionary conclusions about Moses and the Pentateuch and about the chronology of various parts of the *Old Testament*. Undeterred by vehement attacks, he began in 1689 to publish his researches on the New Testament,[3] in which he anticipated to some extent the attempts of the eighteenth and nineteenth centuries to trace the *history* of ancient texts; this 'Text*geschichte*', he believed, must be the basis for an evaluation of the manuscripts and for the constitution of a truly critical text.[4] The point is not that his acute observations anticipated conclusions reached by scholars 200 or 300 years later, but that he applied a new method of establishing a critical text. Yet the unfortunate division[5] between profane and sacred philology prevented Simon's work from having any effect on classical scholarship.

It will always be one of the chief glories of the French Benedictines of the congregation of St. Maur that they made the fundamental editions

---

[1] *Histoire littéraire du sentiment religieux en France depuis la fin des guerres de religion jusqu' à nos jours* (Paris 1916–36, 11 vols. text and 1 vol. indexes), an imposing work of the first rank. Only the first three volumes seem to have been translated into English by K. A. Montgomery (1924–36). Bremond's monograph *Thomas More* (1904) contains some of the best pages ever written on Erasmus.

[2] *The Cambridge History of the Bible*. The West from the Reformation to the present day (1963) 194 f., cf. 218 ff. On Richard Simon see Jean Steinmann, *Richard Simon et les origines de l'exégèse biblique* (1960) and Bruce M. Metzger, *The Text of the New Testament* (2nd ed. 1968) 155 f.

[3] *Histoire critique du texte du Nouveau Testament* (1689).

[4] See on F. A. Wolf below, p. 174; Wilamowi.., '..eschichte des Tragikertextes', *Einleitung in die griechische Tragödie* (1889) pp. 121 ff. Neither Wolf nor Wilamowitz seems to have given any attention to Richard Simon's 'Textgeschichte'; but see the short reference in S. Timpanaro, *La genesi del metodo del Lachmann* (1963) p. 21.1.

[5] Cf. *Philologia Perennis* (1961) p. 13.

of all the Greek and Latin Fathers, a long series of beautiful folio volumes published over a period of more than a century. If the individual parts or volumes were not all of equal value in the use of manuscripts and the constitution of the text, annotation, and the critical distinction between genuine and spurious writings, this was only natural in such a large collection.

In making the edition the Benedictines were in fact carrying out in Catholic France part of the programme of Erasmus.[1] But in Holland too the reformers, above all Grotius, were continuing Erasmus's work in another way.[2] Its universality and greatness became apparent everywhere in the seventeenth century, on the Protestant as well as on the Catholic side.[3]

The greatest among the French Benedictines were Jean Mabillon (1632–1707) and Bernard de Montfaucon (1655–1741). Mabillon was invited to enter the ancient abbey of Saint-Germain-des-Prés[4] and took part in its scholarly work for forty-three years. With his book *De re diplomatica* (1681), written after visits to nearly all the libraries and archives of France, Germany, and Italy, he became the founder of a new branch of scholarship, that of determining the date and genuineness of ancient Latin documents.[5] In his *Traités des études monastiques* (1691, reprinted 1966), in which he justified the labours of the scholar monks, he left us the most beautiful document of Benedictine humanism.[6] Mabillon's confrère Montfaucon[7] laid the foundations of Greek palaeography in his *Palaeographia Graeca* (1708), establishing the principles of the new discipline and providing a list of 11,630 manuscripts. Montfaucon's life's work extended to about fifty folio

[1] See above, p. 78.   [2] See above, p. 127.

[3] Most of the Maurine editions (see Cuthbert Butler's article 'Maurists' in the *Encyclopedia Britannica*) were reprinted in Migne's Greek and Latin *Patrologia* in the nineteenth century, and are still used for those works of which no modern critical texts are available in the new Greek collection of Berlin and in the new Latin one of Vienna. Cf. Ch. de Lama, *Bibliothèque des écrivains de la Congrégation de Saint-Maur* (1882). A. Sicard, *Les études classiques avant la Révolution* (1887).

[4] E. de Broglie, *Mabillon et la société de l'abbaye de Saint-Germain des Prés, 1664–1707* (2 vols., Paris 1888). Henri Leclercq, *Mabillon* (2 vols., Paris 1953–7). M. D. Knowles, *Journal of Ecclesiastical History* 10 (1959) 153 ff.

[5] See the most comprehensive general treatise in palaeography and diplomatics by R. P. Tassin aud Ch. F. Toutain, *Nouveau traité de diplomatique*, 6 vols. Paris 1750–65. Johann Friedrich Böhmer expressly acknowledged that the French Benedictines of St. Maur had anticipated and provided the model for his method of collecting and editing the 'Regesten der deutschen Kaiserurkunden', see F. Schnabel, *Der Ursprung der vaterländischen Studien* (Vortrag 1949, Neudruck 1955) p. 16.

[6] 'Humanitas Benedictina' (1953) = *Ausgewählte Schriften* (1960) p. 180; de Broglie II 295 ff. on Mabillon's labours in old age and especially p. 298 'La pensée . . . est un feu qui fait vivre quand on ne le laisse pas éteindre.' Cf. Petrarch, above p. 16.

[7] E. de Broglie, *Bernard de Montfaucon et les Bernardines, 1715–1750* (2 vols., Paris 1891).

volumes, of which the ten folio volumes of *L'Antiquité expliquée et représentée en figures* (1719) with five volumes of supplements (1724) were the most impressive.

Besides the Benedictines, the Jesuits[1] were active scholars and excellent teachers all through this period. A very able and passionate, though malicious, opponent of Scaliger arose amongst them, Denys Pétau (1583-1652), whose work in the field of chronology we have already noticed.[2] He was also the editor of a complete critical text of Synesius (1612,[2] 1633), which has not yet been completely superseded.

The annotations in the monastic editions became the basis of later commentaries, and more and more learned books amplified the contents of the existing and newly founded libraries in France, now the main models for book collecting in other countries.

A typical example of a successful collector of classical books was the French Protestant Jacques Bongars (1564-1612). Born in Orléans, he studied Greek and Latin in German universities, becoming an admirer of Lipsius in Jena; then he turned to Roman law under Cuiacius in Bourges (1576) and in 1581 in Paris he produced an edition of Justin based on careful collations of new manuscripts, not merely compiled from earlier learned material. From 1586 to 1610, like so many French and Dutch scholars of the age,[3] he was in diplomatic service and saw a good deal of the world from England to Constantinople, inspecting and collecting manuscripts and books and meeting the greatest scholars of the time, with whom he afterwards corresponded: de Thou, Scaliger, Casaubon. He also made copies of inscriptions, the accuracy of which was acknowledged by Mommsen.[4] Bongars did not confine himself to ancient authors,[5] but studied medieval writers as well. Together with his cousin Paulus Petavius (Denys's[6] brother) he was fortunate enough to acquire in 1603 the library of another learned native of Orléans, Pierre Daniel (1530-1603), and of his own teacher Cuiacius. The Bongars collection of about 500 manuscripts and 3,000 printed books, after some wanderings, came to rest in the German part of Switzerland as one of the main treasures of the Berne library[7] (surprisingly perhaps,

---

[1] On Ignatius of Loyola see above, p. 81 n. 2. [2] Cf. above, p. 118.
[3] Cf. Peiresc, below, p. 133. [4] *CIL* v 156.
[5] Virgil, Horace, Ovid, etc. [6] See above, p. 118.
[7] Konrad Müller, 'Jacques Bongars und seine Handschriftensammlung', *Schätze der Burgerbibliothek Bern* (1953) pp. 79-106 with bibliography; Hermann Hagen, *Zur Geschichte der Philologie und der römischen Litteratur* (1879) is important.—A unique item among Bongars's books from P. Daniel's library is the earliest manuscript of the scholia to Ovid's *Ibis*, Cod. Bern. 711, saec. XI. As the few minute leaves could not be found when Hagen arranged the *Catalogus codicum Bernensium* (1875), the 'loss' of the manuscript was—curiously enough—

seeing that it was Geneva with which French scholars had the closest relations).

It seems to have been a pupil of the Jesuits at Avignon, the antiquarian Claude Favre Peiresc (1580–1637), who inspired the French once again[1] to observe, collect, design, and describe the remains of ancient monuments. A great traveller and correspondent, he became a mediator between France, Italy, Holland, and England, and thus a characteristic figure of the time. He was followed by another French traveller and designer of monuments, Jacques Spon[2] of Lyons (1647–85), who is supposed to have been the first to call the branch of scholarship dealing with the monuments of antiquity 'archaeologia' or 'archaeographia'.[3] This was just the time when the great private and public collections originated that were destined to be of the highest importance for the scholars of the eighteenth century.

From a Jesuit college also came one of the greatest lexicographers of all times, Charles Du Cange (1610–88), a great friend of Montfaucon; he had an even harder task than his illustrious predecessors, the Étiennes, as he had to base his *Glossaria ad scriptores mediae et infimae Latinitatis* (1678, 3 volumes) and *Graecitatis* (1688, 2 volumes) on the study of an infinite number of manuscripts, not of printed books. These dictionaries, which were anything but mechanical productions, were reprinted and supplemented through three centuries, and are still unsurpassed. Du Cange was, like Stephanus, also an editor of texts and a historian. His contemporary, Le Nain de Tillemont[4] (1637–98), a pupil of the school of Port-Royal and then a secular priest, wrote two voluminous and immensely learned works. One of them, the *Histoire des empereurs* (from 31 B.C. to A.D. 518) provided Gibbon[5] with the essential

---

lamented again and again without new inquiries in Berne, until in my review of Wilamowitz's *Hellenistische Dichtung*, *DLZ* (1925) p. 2140, I was able to assure the few interested specialists that the codex was in its proper place. F. W. Lenz later used it for his edition of the *Ibis* (1937) pp. 103 ff. (2nd ed. 1956, pp. 129 ff.) and I made use of the many quotations of Callimachus in the scholia, Call. fr. 661 ff. and 789 ff.

[1] See above Italy, pp. 33, 50 ff. On Jesuits and archaeology see A. Rumpf, *Archaeologie* I (1953) 50 f.; ibid. also on forgeries.

[2] *Miscellanea eruditae antiquitatis* I (1679) I. On Spon's various interests and achievements see M. Wegner, *Altertumskunde* (1951) pp. 78 ff. Spon's fellow traveller in the East was a Dutchman, G. Wheler, by one of whose companions drawings of the sculptures of the Parthenon were made in 1674 before their destruction in 1687. Wheler's name is misspelt by Wilamowitz and Rumpf as 'Wheeler' and he is consequently called an Englishman; cf. *Ausgewählte Schriften* (1960) p. 60.

[3] On ἀρχαιολογία, 'antiquitates', see *History* [1] 51.

[4] B Neveu, *Un Historien a l'école de Port Royal, Sébastien Le Nain de Tillemont 1637–1698*, Archives internationales d'histoire des idées 15 (1966).

[5] See below, p. 162.

material for his new concept of Roman history; the other, *Mémoires pour servir à l'histoire ecclésiastique* described the first six centuries of the Church. Tillemont seems to have pioneered the treatment[1] of the political and the ecclesiastical history of the later Roman empire as a single theme. France continued, as we see, to create the tools of research and to lay the foundations for further historical advance, and the seventeenth century was not quite unworthy of the preceding one. But this useful work was mainly done by monks and their pupils in obscurity. Official France, boasting of her own magnitude, had no great estimation of the classics; on the contrary, she grew more and more into the belief that her own literature had surpassed all others, ancient and modern. Charles Perrault in his proud poem 'Le Siècle de Louis le Grand' and in the four volumes of his *Parallèle des anciens et des modernes* (1688–97) finally claimed victory for the *modernes*, that is, of course, the *modernes* of France.[2]

In these discussions Homer, who had been relegated to the background for some time,[3] played an increasing part from the beginning of the seventeenth century onwards, despite Julius Caesar Scaliger's absurd condemnation.[4] After Jehan Samxon's prose translation of the *Iliad* (1530), the series of first-hand translations in verse starts with H. Salel's ten books of the *Iliad* in French (1541, published 1545),[5] to which the rest were added by Amadis Jamyn in 1577. There was a surprising amount of interest, not only in the text of the poems, but also in the traditions about the life of Homer,[6] and it was only in France that anyone was bold enough to use them, as the Abbé d'Aubignac did, to support wild conjectures on the origin of the Homeric epics. All previous translations were surpassed, when in 1711, after long preparations, Mme. Dacier (1654–1720) published her translation of the *Iliad*, followed by the *Odyssey* in 1716; and her renown is still alive,[7] as the latest French translator of Homer assures us. Her father Tanaquil

---

[1] In our time, Eduard Schwartz, a great classical scholar, familiar also with the original sources of church history, took the final step, three centuries after Tillemont.

[2] Cf. above, p. 100. A prehistory of the so-called 'Querelle des anciens et modernes' is given by A. Buck, *Die humanistische Tradition in der Romania* (1968) pp. 75 ff., see esp. p. 90 with supplements to the monographs of H. Rigault (1856) and H. Gillot (1914) on the 'Querelle'. Hans Kortum, *Ch. Perrault und Nicolas Boileau. Der Antike-Streit im Zeitalter der klassischen französischen Literatur* (1966).

[3] See above, p. 104.   [4] Gillot (above, n. 2) p. 204.

[5] Gillot p. 67; cf. Hepp (above, p. 104 n. 3).

[6] Cf. *History* [I] pp. 11, 43, 117.

[7] P. Mazon, *Madame Dacier et les traductions d'Homère en France* (Oxford 1936) esp. pp. 11–13; cf. E. Malcovati, *Madame Dacier, una gentildonna filologa del gran secolo*, Biblioteca del Leonardo 49 (1953).

Lefèvre, who directed her studies, was the very learned[1] editor of many Greek and Latin texts. She followed him in editing a new text of Callimachus (1675), and a number of Latin classics in the series 'in usum Delphini'. The chief editor of this series was Pierre Daniel Huet[2] (1630–1721), the tutor of the Grand Dauphin; he left his books to the Jesuits, and with their library they came into the Bibliothèque nationale. One of Huet's collaborators was the Jesuit Jean Hardouin[3] (1646–1729). He will be for ever notorious for his bizarre theory that most of the Latin classics were productions of the Benedictines in the thirteenth century—a burlesque epilogue to the story of the greatest epoch of French scholarship.

In earlier chapters we followed the progress of humanism and scholarship in Italy and saw how they spread to other countries. In Italy itself there remained an inclination towards virtuosity in Latin verse and prose. This virtuosity is represented by Pietro Bembo[4] (1470–1547), a scholar poet from a Venetian family, made a cardinal in 1539; his Latin works became widely known and a model for others, so that 'Bembismo'[5] became a catchword. But he also wrote studies of Latin grammar and style which did much to maintain the traditional deep love for that language in Italy. In the course of the sixteenth century, however, we find again some remarkable Italians who cultivated pure scholarship; but they were more dependent on the new development of the French Renaissance than on the Italian tradition. For there was not only an interplay between the two nations, but also active collaboration between their scholars. Adrien Turnebus and Henri Étienne on the French, and Francesco Robortello and Piero Vettori on the Italian side, collaborated in restoring the text of Aeschylus.[6] At the same time Denys Lambinus and A. de Muretus worked energetically in the Latin field in Italian libraries, helped by Italian scholars.[7] The outstanding personality of this period was Piero Vettori (Petrus Victorius, 1499–1585);[8] indeed, one is tempted to speak of a 'saeculum Victorianum'[9] in classical scholarship.

---

[1] His excessive learning could even mislead his daughter, see *History* [1] 284 on Call. *Hy.* II 110.
[2] Mark Pattison, *Essays* 1 (1889, repr. 1965) pp. 244–305 'Peter Daniel Huet'.
[3] J. van Ooteghem, *Les Études Classiques* 13 (1945) 222 ff. deals with Hardouin's Horace, but also gives general references to recent literature.
[4] *Opera*, Basle, 1567.
[5] W. Elwert, 'Bembismo, poesia latina e petrarchismo dialettale', *Paideia* 13 (1958) 3–25.
[6] See above, p. 111. on the particulars.
[7] See above, pp. 112 f.
[8] W. Rüdiger, *Petrus Victorius aus Florenz*, Studien zur humanistischen Litteratur Italiens 1
[9] As did Sandys II 135.

Vettori not only finished the edition of the text of Aeschylus, but he also produced a complete edition of Euripides, starting with the *Electra* in 1545, and in 1547 he collated Florentine manuscripts for several plays of Sophocles, among them the great Laurentian manuscript (XXXII.9) which also contained the text of Aeschylus and Apollonius Rhodius. It is characteristic of the leading part now played by France that these editions were printed by the press of the Étiennes in Paris, not by the famous old printing houses of Italy. One of his own manuscripts was a recension of the scholia to the *Iliad* which bear his name as Scholia Victoriana (v).[1] In Greek prose the first place is occupied by his commentaries on Aristotle published between 1548 and 1584, *Rhetoric, Poetics, Politics, Nicomachean Ethics*; indeed he played a part in the rediscovery of the *Poetics* (1560), though he had been preceded by Francesco Robortello (1548). But Vettori's ceaseless activity as editor extended from pre-Hellenistic authors to Dionysius of Halicarnassus (1581), Porphyry (1548), and Clement of Alexandria (1550). His merits as a Latin scholar, especially in his work on Cicero's *Epistulae, Philosophica*, and *Rhetorica*, were no less important. What he could not include in editions and commentaries he collected in twenty-five books of *Variae lectiones*[2] (1553), which he enlarged in later editions (1569 and 1582) to thirty-eight books. Vettori was in correspondence with innumerable contemporary scholars all over Europe. His surviving letters are in the British Museum. All the books and manuscripts that he left came into the possession of the Royal library in Munich;[3] among these the books containing handwritten notes by Politian and many other scholars are of especial value.[4]

Among other members of the Victorian age were Francesco Robortello (1516–67) and Carlo Sigonio (1523–84). Robortello was the first to publish in 1552 the scholia to Aeschylus,[5] which were immediately used by Turnebus[6] for his edition of 1552, as well as by Vettori.[7] He edited, translated, and commented on Aristotle's *Poetics*

---

[1] V is a sixteenth-century Florentine copy of an eleventh-century Athos (?) manuscript, now among the Townleian MSS. in the British Museum (T). Since T was published in full by E. Maass in 1887, the text of V (published by I. Bekker 1827) has no longer any significance except for Victorius's conjectures. See H. Erbse, *Scholia Graeca in Homeri Iliadem* 1 (1969) pp. xxix f.

[2] Cf. above, p. 44 on Politian and p. 111 on Turnebus.

[3] See below on Camerarius's books p. 139.

[4] Handwritten catalogue by Wilhelm Meyer in the Manuscript Department of the Munich State Library.

[5] G. Toffanin, *La fine dell' umanesimo* (1920) pp. 29–45 and *passim*; on the sources of his edition of Aeschylus see R. D. Dawe, *Mnemosyne*, Ser. IV, 14 (1961) 110 ff.

[6] See above, p. 111.   [7] See above, p. 136.

(1548); he was proud of having been the first to print in 1554 the treatise Περὶ ὕψους attributed to Dionysius Longinus until the early nineteenth century, which was often reprinted, translated by Boileau (1674), and passionately discussed especially in France.[1] After these editions and commentaries he published a treatise on the principles of textual criticism, *Disputatio de arte critica corrigendi antiquorum libros* (1557),[2] in which he argued that the critic needs 'iudicium magnum' (as Erasmus's follower, Beatus Rhenanus,[3] had demanded), especially if the manuscripts give no help, as well as 'antiquitatis totius notionem'. In his continual stressing of this point he was in line with J. J. Scaliger[4] whom he could not yet have known. He was rather unfortunate in his quarrel with Carlo Sigonio from 1548 onwards about Roman chronology and Roman antiquities. For in these fields Sigonio[5] was superior to all his predecessors from Glareanus to Robortello, his position, though much more modest, resembling that of J. J. Scaliger in Greek chronology. Sigonio confined himself to writing learned books; unlike contemporary scholars in France and Holland, he was neither a teacher nor an active politician.

Returning to Holland,[6] we find the theme of Robortello's *Disputatio de arte critica*[7] being treated again by Gerard John Vossius[8] (1577–1649) and later by Jean Le Clerc[9] (1657–1736). While Vossius shows himself in his *Aristarchus sive de arte grammatica* (1635 and later) and in his books on the Greek and Latin historians (1624 and 1627) to be mainly a compiler, Le Clerc in his *Ars critica* (1697 and later editions) reveals an acute mind, aware of critical method and able to practise it, as he demonstrated in several editions of Greek classics. This is not surprising when we remember that it was he who made the definitive edition of Erasmus's works in ten volumes[10] which has not been superseded until now.

We have said[11] that the very existence of scholarship depends on the book. After printing had been introduced into Italy in 1465,[12] the first

---

[1] See Sandys II 482 f. s.v. Longinus; see also Jules Brody, *Boileau and Longinus* (1958).

[2] A. Bernardini e Gaetano Righi, *Il concetto di filologia e di cultura classica* (2nd ed. 1953) pp. 46 f. give an adequate appreciation of this book.

[3] See above, p. 84.   [4] See above, p. 118.

[5] Cf. C. Sigonii *Opera* . . . cum notis . . . et eiusdem vita a L. A. Muratorio . . . conscripta (6 vols., 1732–7).

[6] See above, p. 129, where we broke off and dealt with scholarship in France and Italy.

[7] See above, p. 137.   [8] Bernardini (above n. 2) pp. 107–13.

[9] Ibid. pp. 129–46 and S. Timpanaro, *La genesi del metodo del Lachmann* (1963) p. 1 n. 1 and *passim*.

[10] See above, p. 71 n. 1.   [11] *History* [1] 17.

[12] See above, p. 50.

printed Latin and Greek texts and books on classics were produced there; great printing houses were founded in Italy,[1] and in due course in Switzerland[2] and France.[3] Now the Netherlands followed suit when Christopher Plantin's press was established at Antwerp in 1550; inherited by Moretus, it continued to flourish until it was given to the city of Antwerp in 1876 and preserved as a Museum of Printing (Plantin–Moretus Museum).[4] The other great family of Dutch printers was that of the Elzeviers in Leyden and Amsterdam (1580–1712).[5] As scholarship in the Netherlands had a strong tendency to 'polyhistory', an active printing industry of its own was necessary.

A number of classical scholars who lived and worked in the Netherlands were born in Germany or of German descent. They belong to Dutch scholarship and were a link between the two countries, but they could not be called 'Dutchmen'. On the other hand, Janus Gruter[6] was born in Antwerp in 1560, the son of the burgomaster and a highly educated English woman; but after some years in Cambridge and Leyden he spent the greater part of his life in Germany, where he died near Heidelberg in 1627. There he had been appointed librarian in 1602, after he had published his *Corpus inscriptionum antiquarum* with the invaluable assistance of Scaliger.[7] The most depressing experience of his librarianship was in 1623 when a large number of the Palatine manuscripts were presented by Maximilian of Bavaria to the Vatican library. Some of them, including the unique codex Palatinus of the *Anthologia Graeca*, returned to Heidelberg; others went on to Paris.[8] Gruter was himself a poet[9] who wrote hundreds of elegant Dutch sonnets, of which many were translations from Horace and other Latin poets. As a scholar he edited a fair number of Latin prose authors, especially historians, often simply repeating the collected notes of previous commentaries, for instance in his Livy. His own undeniable, if modest, contributions to scholarship were a division of Livy's books into chapters[10] which was generally accepted, and the recognition in a

[1] See above, p. 56.   [2] See above, pp. 83.

[3] See above, pp. 107 ff.

[4] Colin Clair, *Christopher Plantin* (1960) with bibliography and plates.

[5] David W. Davies, *The World of the Elzeviers* (1954). The principal account is by A. Willems, *Les Elzevier* (1880, repr. 1962).

[6] G. Smend, *Jan Gruter* (Bonn 1939); L. Forster, *Janus Gruter's English Years* (1967) with a general bibliography.

[7] See above, p. 116.

[8] See Call. II pp. xcii f. and in general *Handbuch der Bibliothekswissenschaft*, 2. Aufl. III 1 (1955) 576 ff. and 621 f.

[9] See especially Forster (above n. 6) 64 ff.

[10] Cf. above, p. 108 on the division of the text of the Bible into chapters and verses.

Palatine manuscript of a special character of script which is now called Beneventan.[1] In his unlimited productivity and his internationalism he was a truly characteristic figure of his time.

'The progressive influence of the new scholarship in France'[2] upon scholarship in Germany and in other countries was a decisive fact. Melanchthon was an eminent teacher, as we said, and of his many friends and pupils at least two deserve mention: Joachim Camerarius (1500–74) and Hieronymus Wolf (1516–80). They were teachers of distinction and heads of the newly founded Protestant schools in Nuremberg and Augsburg respectively, but they were also scholars superior to Melanchthon, and both were great editors. Camerarius's[3] edition (1535) of Ptolemy's *Tetrabiblos*, a most important astrological work, with a Latin translation by Melanchthon, has not yet been completely superseded. In 1538 Camerarius and Grynaeus published the first Greek edition of Ptolemy's astronomical work Μεγάλη σύνταξις, the *Almagest*.[4] In the long series of his other Greek and Latin texts the foremost was his Plautus of 1552 for which he was able to use two new manuscripts, the Palatini B (Codex Vetus Camerarii) and C (Codex alter Camerarii decurtatus). Nicholas of Cusa had brought a manuscript with twelve new comedies to Rome (Vat. D) as early as 1429,[5] but it was only now that, thanks to Codex B, the text of all the extant plays was complete. Camerarius possessed a very wide knowledge of the ancient world, akin to the learned encyclopedism of the seventeenth century, but still more cultured, sympathetic, and human. All his extant manuscripts and letters are, like the books and manuscripts of Piero Vettori, in the Bavarian State Library, the 'Camerariana'.[6]

Hieronymus Wolf was a pupil of both Camerarius and Melanchthon.[7] Born in Oettingen, he left to posterity a gloomy account of his life; this was appreciated by another important and equally unhappy scholar,

---

[1] See E. A. Lowe, *The Beneventan Script* (Oxford 1914).   [2] See above, p. 94.

[3] F. Stählin, 'Camerarius', *Neue Deutsche Biographie* vol. 3 (1957) 104 f.

[4] On Regiomontanus and Bessarion see above, p. 37 n. 8.

[5] F. Ritschl, *Opuscula philologica* II (1868) 5 ff. On Nicholas of Cusa (called also Nicolaus Treverensis), secretary to Cardinal Orsini, as discoverer and collector of classical manuscripts see Frank Baron, 'Plautus und die deutschen Frühhumanisten', *Studia humanitatis, Ernesto Grassi zum 70. Geburtstag* (Humanistische Bibliothek, Abhandlungen und Texte, Reihe 1: Abhandlungen, vol. 16 [1973]) pp. 89–101.

[6] Clm 10351–14431; cf. K. Halm, *Über die handschriftliche Sammlung der Camerarii und ihre Schicksale* (1873).

[7] G. C. Mezger, *Memoria Hieronymi Wolfii* (1862); H.-G. Beck, 'Hieronymus Wolf', *Lebensbilder aus dem bayerischen Schwaben* 9 (1966) 169–93; cf. 'Augsburger Humanisten und Philologen', *Gymnasium* 71 (1964) 201 f. A number of books, once part of Hieronymus Wolf's library, were recently identified by Heinz Dollinger (1973) who will publish an article about them.

J. J. Reiske, who appropriately printed it in his *Oratores Graeci*.[1] For Wolf was the most famous editor and translator of the Attic orators. His Isocrates appeared in 1548, Demosthenes in 1549, and the definitive edition of 1572 in six volumes with scholia and annotations remained fundamental for more than two centuries. The codex Augustanus (A, now Monacensis graec. 485) was Wolf's favourite manuscript, naturally enough, seeing that it was in Augsburg that after years of restless wanderings he found a place of permanent residence, first as Johann Jacob Fugger's secretary and librarian (1551–7),[2] and then as headmaster of the school of St. Anna and administrator of the city library (1557–80). For his Greek studies he would have preferred Paris, where they had then just reached their peak, if as a heretic he had not been afraid of persecution by the Sorbonne. In Augsburg,[3] however, the rich merchant house of the Fuggers had built up a remarkable library of classical and Byzantine authors, not without help from Henri Étienne in Paris, who was not only a great scholar and printer, but also a book collector.[4] In 1571 the early printed books and the manuscripts (about 180 Greek and many Hebrew ones) were bought by Duke Albrecht of Bavaria for his Munich Palace library, and more than three centuries later it was discovered[5] that a great number of them still had their shelfmarks entered in Wolf's own hand, so that the original imposing Fugger library could be reconstructed. As an editor Wolf went far beyond the Attic orators; in 1557 he initiated a gigantic enterprise by publishing the first editions of Ioannes Zonaras, Nicetas Choniates, and Nicephoros Gregoras as part of an 'integrum Byzantinae historiae corpus'.[6] He was indeed the founder of modern Byzantine scholarship.

His contemporary Wilhelm Holtzmann (Xylander, 1532–76), born in Augsburg, and professor and librarian in Heidelberg from 1558, edited both classical and Byzantine prose authors.[7] He is best known as

---

[1] *Oratores Graeci*, ed. J. J. Reiske VIII (1773) 772–876.

[2] Paul Lehmann, *Eine Geschichte der alten Fuggerbibliotheken* I (1956) 31 ff.

[3] Among his many visitors from abroad was Roger Ascham, see below, p. 143.

[4] See above, pp. 109 ff. Cf. P. Lehmann (above n. 2) pp. 81 ff. and *passim*.

[5] Otto Hartig, 'Die Gründung der Münchener Hofbibliothek durch Albrecht V und J. J. Fugger', *Abhandlungen der Bayerischen Akademie der Wissenschaften*, Philos.-philol. u. hist. Klasse XXVIII 3 (1917).

[6] F. Husner, 'Die editio princeps des "Corpus historiae Byzantinae"', *Festschrift Karl Schwarber* (1949) pp. 143 ff.

[7] On Xylander's Plutarch see Sturel (above, p. 113 n. 1) pp. 440 ff. and Aulotte (ibid.) pp. 31–4. The *editio princeps* of Marcus Aurelius (1559) was the work of Conrad Gesner, not of Xylander, as Sandys II 270 and others wrongly say; see Marc. Aurel. ed. Farquharson I (1944) xxii ff.

a translator and editor of Plutarch (Greek text 1572), because the page numbers of the 1599 impression of his edition are still used in references to the *Moralia*. We should remember that at the same time Amyot's translation made Plutarch a French classic.[1]

An able and industrious scholar, Friedrich Sylburg (1536–96) completed the short-lived Xylander's edition of Pausanias (1583) and followed it with a series of his own editions. While employed both as a press reader to the great printers and publishers Wechel in Frankfurt and Commelinus in Heidelberg and as a sublibrarian in Heidelberg he managed to publish Greek and Latin classics and patristic texts with great care and success.[2]

Xylander followed Wolf as an editor of Byzantine writers only in a modest way, confining himself to Cedrenus and Psellus. But Wolf's pupil David Hoeschel[3](1556–1617, headmaster of St. Anna from 1593) was more productive. His first edition of Photius' *Bibliotheca* (1601), based on four good manuscripts, was the means of preserving the patriarch's unique excerpts from 280 subsequently lost Greek works. He was especially fortunate in having the assistance of J. J. Scaliger with corrections and additions, not only for his Photius, but also for his first editions of Procopius (1607) and Phrynichus (1601). The foundation of a printing house for these and other editions was financed by the family of the Welsers; the most learned of them, Marcus Welser, a correspondent of Scaliger and a pupil of Muretus in Rome, published in Antwerp in 1598 at least a part of the Roman *itinerarium* discovered by Celtis.[4] The map was finally called *Tabula Peutingeriana* in the edition of 1618, a name hardly deserved, but generally accepted. With the financial help of the Fuggers a Jesuit College was founded in 1582 of which the luminary was Jacobus Pontanus, not only an elegant Latin stylist, but also like his colleagues in the Protestant school of St. Anna, an editor of Byzantine historical and theological works.

It is evident from this short list of names that there were no longer German scholars of European distinction.[5] Scholarship lived quietly on

---

[1] All the translations of Plutarch into modern languages are indebted to the earliest Latin version of Bruni (see above, p. 29).

[2] Surprisingly well informed is the treatment of Sylburg in B. A. Müller's review of W. Kroll, *Geschichte der klass. Philologie* (2nd ed. 1919), *Philolog. Wochenschrift* 46 (1926) pp. 1164 ff. See also K. Preisendanz, 'Aus F. Sylburgs Heidelberger Zeit', *Neue Heidelberger Jahrbücher*, N.F. (1937) pp. 55–77.

[3] 'Augsburger Humanisten und Philologen', *Gymnasium* 71 (1964) 203 f.

[4] See above, p. 64.

[5] Some had emigrated to Holland where they hoped to find peace and better conditions for scholarly work.

in schools and universities, in printing offices and libraries.[1] It dwindled, but it survived, and not even the Thirty Years' War confronted it with such a deadly crisis as had occurred at the end of antiquity.

[1] For details see C. Bursian, *Geschichte der classischen Philologie in Deutschland* (1883) pp. 219–356.

XI

RICHARD BENTLEY AND CLASSICAL SCHOLARSHIP IN ENGLAND

IN our survey of the seventeenth century we found that the characteristic scene in most Continental countries was that of self-satisfied polymaths filling enormous volumes with collected antiquities and reproducing in their editions of texts the accumulated notes of the last two centuries. There were also some individual scholars slowly working for themselves on the traditional lines. But never in our history was stimulating criticism more urgently needed than at that moment; and at exactly that moment there arose in England the critical genius of Bentley.

Richard Bentley (1662–1742) was the last product of the seventeenth century. Classical scholarship was nothing new to England. In nearly every chapter from Poggio's time on, we have had occasion to consider the relation of classical studies in England to those in the leading Continental countries.[1] But never before had they risen to the highest level. They were at their liveliest in Erasmus's time. But this was not because the circle of his noble British friends formed a group of industrious scholars such as we found elsewhere; the fact was simply that the royal family, the nobility, and the clergy maintained the Christian–humanistic tradition even after the separation from Rome. General classical education was promoted, and Roger Ascham's[2] (1515–68) *Schoolmaster* illustrates how it flourished, especially in Cambridge. He tells us how he found Lady Jane Gray reading and enjoying her Plato; as private tutor to Elizabeth before and after she became queen, he read Sophocles, Isocrates, and the Greek New Testament with her. He was a great traveller, and stayed several years in Germany at Augsburg with Hieronymus Wolf;[3] but in Italy he found humanism declining.

An immense stock of translations[4] began to grow up, enriching the

---

[1] See above, pp. 61 f. (Poggio, Enea Silvio); 65 (Th. Linacre); 72 (Colet); 73 (Erasmus); 79 (Th. More).

[2] R. Ascham, *English Works* (*Toxophilus; Report of the Affairs and State of Germany; The Schoolmaster*) ed. William Aldis Wright (1904, repr. 1970). Cf. also M. L. Clarke, *Classical Education in Britain, 1500–1900* (1959).

[3] See above, p. 140. [4] Sandys II 239 ff.

English language and its literature. George Buchanan and others wrote better Latin poems than anybody on the Continent in the second half of the sixteenth century. There was also a line of scholars in the strict sense which began just after the end of the Elizabethan age. Henry Savile, Provost of Eton and Warden of Merton College Oxford until 1622, the host of Casaubon, produced the fundamental edition of John Chrysostom in eight magnificent folio volumes. He inspired the next generation, of whom we may mention John Selden,[1] John Hales, and Thomas Gataker,[2] and in the second half of the seventeenth century John Pearson,[3] bishop of Chester, Thomas Stanley,[4] the editor of Aeschylus and Bentley's rival as editor of Callimachus, Gale, Potter, and Barnes. They were scholars of considerable achievement, especially in Greek, and equal, if not superior, to their contemporaries in all other countries.

It is, however, the continuance of Christian Platonic humanism that is especially characteristic of England. We find it in Spenser's poetry as well as in the philosophy of the Cambridge Platonists. It takes the middle road, like the Arminians[5] in Holland, regulative and tolerant, opposed on the one side to puritan fanaticism, on the other to deism and free-thinking. The best of the scholars mentioned above were in this tradition: John Selden, for instance, in his tabletalk demonstrated the noble superiority of classical Christian education, and John Hales imbued even his contributors to theological controversy with charm and humanity. If at first surprising, it is quite logical that at the end of the seventeenth century the most powerful defenders of the classical, as well as of the Christian, tradition in the struggle against deists and free-thinkers were three British and Irish clergymen, very different one from another, but united in this common cause—the greatest satirist, Swift, the greatest philosopher, Berkeley, and the greatest classical scholar, Bentley.

We observed that men of religious personality, such as Scaliger, had never discussed the relation of Christianity to the ancient world as a *problem*. It was the same with Bentley; with the absolute firmness of a long tradition he embraced both alike in his life and in his writings. A new feature[6] in his make-up was the influence of contemporary science then becoming dominant; he was well acquainted with

---

[1] See above, p. 119.

[2] See Marc Aurel, ed. Farquharson (above, p. 140 n. 7) 1 pp. xlv ff. and G. Zuntz's just appraisal of Gataker's commentary (1652) in *Journ. of Theol. Stud.* 47 (1946) 85.

[3] See E. Fraenkel, Aesch. *Ag.* 1, pp. 78 ff.

[4] See below, p. 152; cf. Call. II, pp. xliv f. and *History* [1] 134.6, 281.

[5] See above, p. 124.

[6] On science and scholarship in France see above, p. 114 n. 4 (J. J. Scaliger and Pierre de la Ramée).

Newton's publications, and became his personal friend, as we shall see.[1]

Born in 1662, Bentley lived to the age of eighty and most of his writings appeared after 1700. But he had the incredible vitality of the seventeenth century, sometimes, it might seem, to an excessive degree. The faculty for immense reading, the almost unlimited knowledge of languages and subject-matter were characteristic of this period. But at the same time we recognize a new refinement of spirit, an absolute confidence in the power of reason for analysing and criticizing tradition and for finding the legitimate order in the creations of the human mind. These were tendencies of the new age.[2] Bentley belongs to both periods, and that may perhaps explain some of his strange or even bizarre behaviour. But behind all the external disharmonies, there was a firm self-conscious personality.

Bentley was born in a small town in Yorkshire, Oulton near Wakefield, was taught Latin by his mother, and sent to the Wakefield grammar school; at the age of fourteen he entered St. John's College, Cambridge, and took his first degree at eighteen. But he could not get a fellowship at St. John's, since two Yorkshiremen already held such fellowships. So after holding a mastership at Spalding school for a short time he became private tutor to the son of the Dean of St. Paul's, Dr. Stillingfleet, and lived for six years with the Dean and his family in London. Stillingfleet had one of the finest private libraries of the time, and the free use of this library laid the solid foundation of Bentley's wide learning in classics, theology, philosophy, and science. In 1690 Bentley was ordained, Stillingfleet was made Bishop[3] of Worcester, and Bentley went with his pupil to Wadham College, Oxford. He immensely enjoyed the treasures of the Bodleian Library, which, as we are told,[4] gave him a preference for Oxford over Cambridge, and made possible his first literary projects. But at the end of the same year he had to leave Oxford to take up residence with the Bishop of Worcester. It was from Worcester that he wrote his *Epistula ad Millium*, that is, to his friend Dr. John Mill, Principal of St. Edmund Hall, Oxford.[5] The publication of this 'letter' in 1691 made his reputation as a classical scholar.

---

[1] Below, p. 147.   [2] They go far beyond Valla's rationalism (cf. above, pp. 41 and 75 f.)

[3] Bentley might have been amused to know that he would appear as 'the very orthodox . . . future Bishop of Worcester' in modern American and German literature, see A. Koyré, *From the Closed World to the Infinite Universe* (Baltimore, Md. 1957) p. 189, taken over by F. Wagner, 'Neue Diskussionen über Newtons Wissenschaftsbegriff', *Sitz. Ber. Bayer. Akad. der Wissenschaften, Philos.-histor. Klasse,* Jahrg. 1968, Heft 4, p. 14.

[4] Monk 1 18n. (below, p. 148 n. 8).

[5] *Epistula* reprinted with introduction and notes by G. P. Goold (1962). A. Fox, *John Mill and Richard Bentley* (1954); cf. below, pp. 149 f.

In the next year we find Bentley in another field. Robert Boyle, one of the foremost scientists of the age, who was convinced that reason and religion could be reconciled, had just died and left fifty pounds for the preacher of eight sermons in which the Christian religion was to be vindicated against notorious infidels, such as atheists and deists. The trustees appointed Bentley, who from the pulpit of St. Martin's church in London delivered his 'Boyle lectures' from March to December 1692, 'A confutation of atheism', in which he tried especially to reveal Thomas Hobbes as an atheist in the disguise of a deist. Bentley did not appeal like a theologian to the authority of sacred books, but, as he said, 'to the mighty volumes of visible nature and the everlasting tables of right reason'. In the last three lectures he took up Newton's great discoveries, published in his *Principia* five years before, and used them to prove the existence of an intelligent and omnipotent creator. The Boyle lectures gave rise to the famous *Four letters from Sir Isaac Newton*.[1] Newton was highly pleased and approved Bentley's arguments in general, pointing out just a few mathematical fallacies. The whole of Bentley is to be seen in the Boyle lectures:[2] he is the perfect controversialist, not declaiming passages of the scriptures, but proving his point by proper arguments. He is also able to develop an impressive eloquence; thus in the peroration of the last of the eight sermons he made the following comparison:[3]

We have formerly demonstrated[4] that the body of a man, which consists of an incomprehensible variety of parts, all admirably fitted to their peculiar functions and the conservation of the whole, could no more be formed fortuitously than the *Aeneis* of Virgil, or any other longer poem with good sense and just measures, could be composed by the casual combinations of letters. Now, to pursue this comparison,[5] as it is utterly impossible to be believed, that such a poem may have been eternal, transcribed from copy to copy without any first author or original; so it is equally incredible and impossible that the fabric of human bodies, which has such excellent and divine artifice and, if I may so say, such good sense and true syntax, and

[1] R. Bentley, *Works* ed. A. Dyce III (1838, repr. 1966) 201 ff.

[2] The first to draw my attention to the sermons was my friend the late Professor of English in the University of Hamburg, Emil Wolff, to whom I owe more in the whole chapter on England than I can now remember.

[3] R. Bentley, *Works* III 200.

[4] In the fifth sermon, ibid. pp. 112 ff.

[5] I have often wondered whether 'this comparison' was invented by Bentley himself; the nearest parallel I have been able to find, which may have suggested to him his own version, seems to be Augustine, *Civ. dei* XI 18 'deus . . . ita ordinem saeculorum tamquam pulcherrimum carmen . . .' See E. R. Curtius, *Europäische Literatur und lateinisches Mittelalter* (1948) pp. 401 ff. and 441 ff.

harmonious measures in its constitution, should be propagated and transcribed from father to son without a first parent and creator of it . . .[1]

The striking point in this argument is that for Bentley the natural example of perfect teleology, suitable and reasonable as a whole and in its parts, is the great classical poem; the analogy is so strict that the human organism even has its grammatical and metrical qualities, true syntax and harmonious measures like poetry, and the fabric of human bodies is a transcription from father to son like the copies of a text. In both cases there must be a first author and an original. The important point for classical scholarship behind Bentley's line of argument is the belief in the original harmony of classical poetry, its good sense and just measures which—if corrupted by transcription from copy to copy— must be restored by reasonable criticism. So these Boyle lectures reveal a characteristic blend of his Christian theology with his humanistic scholarship and his firm grasp of principles.

The lectures, as soon as they were published, and also through translation in Holland and Germany, made a deep impression; but his second series of Boyle lectures, 'A defence of Christianity', delivered two years later, were never printed and seem to have been lost. In the meantime Bentley, still residing with the Bishop of Worcester, was appointed keeper of the Royal libraries in 1694, then Chaplain to the King in 1695, and in 1696 he took possession of the rooms for the Royal Librarian in St. James's Palace, next to the Earl of Marlborough, the future hero of Blenheim; in the same year he took the degree of Doctor of Divinity at Cambridge, and preaching before the university again defended Christianity against deism. From a letter of 1697[2] we learn that a circle met every week in Bentley's lodgings in St. James's consisting of his truest friends, John Evelyn (who in vain urged him to publish the second course of his Boyle lectures), Christopher Wren, John Locke, and Isaac Newton.[3] This was the year in which the first *Dissertation upon the Epistles of Phalaris* appeared as an appendix to a book by his friend Wotton.[4] The enlarged edition of 1699 established Bentley's reputation throughout Europe.

It happened that towards the end of this year the mastership of Trinity College Cambridge became vacant. The commission appointed

---

[1] The quotation is deservedly put in the text because of the importance of its content. It is also a good example of Bentley's style—though to appreciate it fully one ought to read right on to the concluding 'Amen'.

[2] Bentley, *Correspondence* I (1842) 152 (21 Oct. 1697).

[3] Cf. above, p. 145 n. 3.

[4] William Wotton, *Reflections upon the Ancient and Modern Learning* (1694, 2nd ed. 1697).

by the King to fill it unanimously recommended Bentley, and he was installed in 1700 at the age of thirty-eight. He married in the following year, when he was also Vice-Chancellor of the university. In 1717 he was elected Regius Professor of Divinity, actually by his own vote through an extraordinary manipulation.[1] In the theological field he had published his *Remarks*[2] *upon a late Discourse of Freethinking* [*of Anthony Collins*][3] in 1713, and had delivered his *Sermon upon Popery*[4] in 1715 and his prelection on the three heavenly witnesses in 1 John 5:7 in 1717. In the classical field he published an edition of a text from time to time, generally at a moment when he needed the help of a mighty patron. For the forty-two years of his mastership of Trinity[5] until his death in 1742 were anything but a golden age of peace for the college and the university.[6] Bentley was by no means a dispassionate administrator, but an imperious ruler in continuous conflict with the Fellows of Trinity and other members of the university. The episodes of this Homeric struggle are thrilling indeed and almost unique in the honourable history of classical scholarship;[7] but they are recounted in all the descriptions of Bentley's life.[8]

[1] Monk (below, n. 8) II 8 ff.   [2] *Works* III 287 ff.

[3] Collins's *Discourse* reprinted, translated, and introduced by G. Gawlick, Stuttgart 1965 (La Philosophie et communauté mondiale [2]).

[4] *Works* III 241 ff.

[5] Bentley was anxious to promote scientific studies, mathematics, astronomy, and chemistry among the members of his college.

[6] In contrast, his family life seems to have been peaceful and happy.

[7] There was at least no threat of bloodshed, as in the conflict between Poggio and Valla, see above, p. 34.

[8] There is no satisfactory modern biography of Bentley. The standard work is still J. H. Monk, *The Life of Richard Bentley* (1830, 4°; second edition, revised and corrected, 2 vols., 8°, 1833, reprinted 1969); Monk was a classical scholar, who became Regius Professor at Cambridge in 1806 and subsequently Bishop of Gloucester. A still more competent and elegant, but short book is R. C. Jebb's *Bentley* (English Men of Letters, 1882, reprinted 1968); Jebb is less concerned with biographical details than with Bentley's scholarship and with his style as a great English writer. In contrast to Jebb the author of *Dr. Bentley. A study in Academic scarlet* (London 1965), R. J. White, seems to be well acquainted as a historian with the age in which Bentley lived, but to lack the qualifications for writing a full-length book on a scholar whose eminence lay in his work on Greek and Roman classics.—Of the many articles on Bentley the most valuable seems to me to be Friedrich August Wolf's sketch of a life of Bentley in his *Literarische Analekten* I 1–89 and I 493–9 (1816, reprinted in his *Kleine Schriften* II, 1869, 1030–94, see below p. 173 n. 1). The collection of Bentley's *Works* edited by Alexander Dyce (1836–8, repr. 1966) was discontinued after the third volume and contains only a small fraction of his life's work. His *Correspondence* was edited by Christopher Wordsworth (2 vols., 1842). See also J. Bernays, 'R. Bentley's Briefwechsel', *Rh.M.* N.F. 8 (1853) 1 ff. There are also some editions of letters to special persons. Many of the most important notes from the margins of his books have been published in various articles, books, and editions; the references are to be found in A. T. Bartholomew, *R.B., a bibliography* (Cambridge 1908). In honour of the tercentenary of Bentley's birth a fair number of articles and speeches were published in 1962. See especially *Proceedings of the Classical Association* 59 (1962) 25 ff., where J. A. Davison reminds us (p. 34) of Bentley's still uncollected works.

We shall deal first with Bentley's two outstanding earlier publications, the *Epistula ad Millium* and the *Dissertation upon the Epistles of Phalaris*, then with his editions of ancient texts and fragments, and finally with those great projects which he could not carry out himself.

The only manuscript of the chronicle, the χρονογραφία, of Iohannes Malalas (written in the second half of the sixth century A.D. in Antioch in Syria) is in the Bodleian Library. Scaliger's work had created an interest in all the remains of chronologists, and some Oxford scholars had quoted the codex; two had even prepared an edition, but had died before printing started. So in 1690 Humphrey Hody, the college tutor of Bentley's pupil Stillingfleet, was entrusted with its editing and John Mill with the general supervision. The proof sheets were sent to Bentley, who was asked by his friend Mill to communicate any suggestions that might occur to him. Bentley's suggestions in the form of a letter to his friend were published as an appendix to Hody's edition of the chronicle, filling 98 pages. There are a number of quotations from earlier Greek authors in Malalas' chronicle; Bentley emended corrupt passages with the help of texts still unpublished but known to him from Oxford manuscripts in the Bodleian[1] or in college libraries.

He was mainly interested in references to the Attic dramatists. For instance when the chronicle referred to a play by Euripides about Pasiphae[2] Bentley (without the help of our modern collections and bibliographies) was able to state that there was no play of Euripides called Pasiphae, but that the story was the subject of Euripides' Κρῆτες. He demonstrated that an anapaestic fragment (fr. 472 N.²) of Euripides quoted by Porphyrius belonged to the Κρῆτες, not, as Grotius[3] had said in his *Excerpts from Greek Dramatists*, to the Κρῆσσαι; *en passant*, he emended the corrupt text and then explained the structure of the anapaestic dimeter, the essential law of which had been ignored by all modern scholars and neglected in all modern imitations. Grotius for instance, in his translation of this fragment, supposing like everybody else (including Scaliger and Buchanan) that the last syllable of every dimeter was *anceps*, had often put a tribrach at the end instead of an anapaest. But Bentley observed that the anapaests ran on to the

---

[1] For instance Theognost's *Orthography*, with many ancient quotations was discovered by Bentley (first published in 1835 in Cramer's *Anecdota Graeca Oxon.* II 1–165).

[2] Ioh. Malalas p. 86.10 ed. Dindorf (1831) περὶ δὲ τῆς Πασιφάης ἐξέθετο δρᾶμα Εὐριπίδης ὁ ποιητής Eur. fr. 471 f. N.², H. v. Arnim, *Supplementum Euripideum* (1913) pp. 22 ff. (Κρῆτες = *Berliner Klassikertexte* V 2.73. *TGF* fr. 472 a = *P. Oxy.* 2461, fr. 1.12, *Supplementum ad TGF²* ed. B. Snell (1964) p. 9.

[3] See above, p. 127.

paroemiac as if the whole had been one continuous verse, and thus discovered the metrical continuity of the anapaestic system. From Greek metre he turned to the anapaestic fragments of Latin tragedy, put them right, and even discovered a new one in a poetical quotation of Cicero.[1] A long quotation from Sophocles[2] in several Christian writers he showed to be spurious by proving that one of its expressions was a Hebraism which occurred in the Old Testament (line 4 καρδίᾳ πλανώμενοι).

In another passage the chronicle gives the name Minos to one of the dramatists listed in its text; Bentley pointed out that not Minos, but Ion of Chios[3] was meant and added an exhaustive discussion of the man and his works. He was completely familiar also with the fragments of Hellenistic poetry and with all the lexicographers from Hesychius to Suidas and the Etymologies. The whole work is charged with an exciting exuberance of knowledge of language, subject-matter, metre, literary history, and a style that is colloquial, lively, persuasive, and full of humour.

The *Dissertation upon the Epistles of Phalaris* is a work of the same character, but written in English instead of Latin, an innovation that marks an epoch in classical scholarship.[4] The origin of the work calls for some explanation.

At the end of the seventeenth century the French mind had claimed the superiority of the moderns over the ancients;[5] the controversy which ensued found its way to England and was the starting-point for a debate on the Epistles of Phalaris. For Sir William Temple,[6] the illustrious statesman, defending the ancients in an 'Essay upon ancient and modern learning' (1692), rather surprisingly expressed a strong admiration for Aesop's fables and Phalaris' epistles: 'I think the epistles of Phalaris to have more race, more spirit, more force of wit and genius than any others I have ever seen, either ancient or modern.'[7] This advertisement caused a sudden demand for the ancient thriller, and the Dean of Christ Church, Oxford, encouraged a young gentleman of great promise in his college, the Honourable Charles Boyle, grand-nephew of Robert Boyle,[8] to make a new edition, which was published in 1695.

---

[1] *Works* II 276.   [2] *Works* II 256 ff. and Addenda pp. 357 f. = [Soph.] fr. 1126 Pearson.

[3] *Works* II 304 ff.; on Bentley's emendations of many of Ion's fragments see *TGF*² pp. 732 ff.

[4] Cf. above, p. 110. Henricus Stephanus had written in French, but only in special circumstances.

[5] See above, p. 134.

[6] On W. Temple see H. W. Garrod, 'Phalaris and Phalarism', in *The Study of Good Letters* (1963) pp. 123 ff.; Swift was 'servant' to Temple. See also K. Borinski, 'Die Antike in Poetik und Kunsttheorie' II = *Erbe der Alten* 10 (1924, repr. 1965) 104 ff. 'Der Ritter Temple'.

[7] *Works*, new ed. vol. III (1841, repr. 1968) p. 478.   [8] See above, p. 146.

Meanwhile Bentley's friend, William Wotton, had subjected Temple's essay to a calm examination in his 'Reflections upon ancient and modern learning', in which acting as a sort of mediator he pointed out that the ancients were superior in eloquence and poetry, the moderns in science—which Temple had completely disregarded.

Bentley[1] himself had said in private conversation that the 'Epistles' were a spurious piece, unworthy of a new edition, and when in 1697 a new edition of Wotton's 'Reflections' was called for, he hastened to contribute an appendix entitled 'A dissertation upon the Epistles of Phalaris'. He had been infuriated by an ironical gibe in Boyle's preface to the effect that Bentley 'out of his singular humanity' had denied him the necessary time for collating the manuscript of the Epistles in the King's Library. Bentley now told the true facts of the story, concluding it with the bitter remark: 'They [Boyle and his friends] ought to have made inquiries before they ventured to print, which is a sword in the hand of a child.' The phrase 'a sword in the hand of a child' was not coined by Bentley, but an adaptation of the Greek proverb μὴ παιδὶ μάχαιραν,[2] most appropriate in this place, and a good example of his allusive style.[3] In its second and final edition Bentley's 'Dissertation' was transformed from a modest appendix of 98 pages into a large volume of about 600 pages; and the author, who had started in a defensive position, came in the heat of the polemics[4] to play a more and more aggressive part. As Bentley was now writing in English,[5] his style became more explosive than in the Latin *Epistula* and shocked many of his readers; as a writer indeed he was not unequal to the greatest English satirists of his age. In the lengthy controversy Swift, who was in the service of Sir William Temple,[6] naturally fought on the side of Bentley's adversaries in his *Tale of a Tub* and in his *Battle of Books*. One short passage must be quoted from the fable of the bee and the spider in the *Battle of the Books*. Swift[7] lets the bee say against the spider: 'The difference is, that, instead of dirt and poison, we have rather chosen to fill our hives with honey and wax, thus furnishing mankind with the

---

[1] *Works* I p. ii.

[2] References are given in *Corpus paroemiographorum Graecorum*, edd. Leutsch-Schneidewin (1839) on Ps.—Diogenian. VI 46 (not in the codex Athous).

[3] It would be good to have a reliable text of Bentley's writings with references to his quotations and literary allusions.

[4] A complete bibliography of the pamphlets from both sides is given by Woldemar Ribbeck in his German translation of the 'Dissertation' (Leipzig 1857) pp. xxviii–xxxii.

[5] See above, p. 150.

[6] See above, p. 150 n. 6.

[7] Swift, *Prose Works*, ed. Temple and Scott (1907) I 172; cf. Borinski (above, p. 150 n. 6) p. 105.

two noblest of things, which are sweetness and light.' 'Dirt and poison' refers to Bentley, and this pungent phrase shows the impression which his writing made on his contemporaries. Where style was concerned there were dirt and poison on both sides, *pace* Swift; but with regard to the facts of the matter Bentley was absolutely in the right, and the final victory was his.

Bentley's dissertation upon the letters of Phalaris is the most surprising product of the 'Querelle des anciens et modernes'. Temple and all those who followed him were on the side of the 'ancients', that is, the writers whom they regarded as the best representatives of ancient literature. Must we say then that Bentley, because he opposed them, was therefore an adherent of the moderns, a conclusion that might seem to be supported by his friendship with scientists like Newton? There appears to be no evidence for this view in Bentley's own writings. His main concern in the 'Dissertation' was to examine the Phalaris letters exhaustively and methodically and to provide irrefutable proof of the spuriousness of these 'declamatiunculae', as Erasmus had rightly called them.[1] In doing so he had to investigate questions of chronology and especially of language. He had also to deal with the age of Pythagoras, with the history of Sicily, in particular Zancle and Messene, and with Sicilian money. But at the centre of his argument there was again, as in the *Epistula*, the great Attic literature of the age of tragedy, comedy, and satyr-play, with a fundamental chapter on Attic dialect and atticistic imitations. To a still greater extent than in the earlier book he was able to display the full power and maturity and the stupendous range of his scholarship.

It seems almost incredible that in the years between the two editions of the 'Dissertation' Bentley was able to present Graevius[2] with his new collection of the fragments of Callimachus, which he had first promised in 1693 and finished for the printer in 1697.[3] But this work of pure learning led to recrimination,[4] for Bentley was alleged to have used the notes of Thomas Stanley[5] without acknowledgement. In fact, however, he had never seen Stanley's diligent collection of about 250 Callimachean

---

[1] Bentley, *Works* 1 80 (Phalaris): 'The great Erasmus . . . his words: "those Epistles . . . what else can they be reckoned than little poor declamations?" ' They should not be called forgeries, as e.g. by G. N. Clark *The Seventeenth Century* (1948²) p. 271, because nobody in antiquity would have been deceived by them.

[2] One of the scholars born in Germany, but working in Holland (see above, p. 138).

[3] See *Correspondence* (above, p. 148 n. 8) 1 53, *passim*. A. C. Clark, 'Die Handschriften des Graevius', *Neue Heidelberger Jahrbücher* 1 (1891) 238 ff.

[4] See Call. II pp. xliv f. for details of the supposed fraud.

[5] See also Thomas Stanley, *The Poems and Translations* ed. G. M. Crump (1962).

fragments from printed sources, respectable achievement though it was, seeing that his predecessors, Vulcanius and Anna Fabri, had discovered respectively no more than 86 and 53 fragments altogether. It was Bentley's unique knowledge of the manuscripts[1] that enabled him to make great advances; not only was the number of identified quotations from Callimachus raised to 417, but their corrupt text was in many cases emended, partly from new manuscripts but mostly by Bentley's emendatory genius.[2] Many even of his boldest conjectures have been completely confirmed by the papyri.[3] On the other hand, his authority was such that when he went wrong he almost inevitably led later editors astray.[4] Henricus Stephanus[5] had once tried to make a collection of the Greek lyric fragments. But Bentley's Callimachus was the first methodical work in this field; the collection itself was exhaustive for its time, and by attempting to arrange the fragments in the order of the lost works and to give some cautious reconstructions he made it exemplary for posterity. Though he contributed emendations to about 300 fragments in Leclerc's[6] edition of Menander and Philemon, he made no further collection himself after the Callimachus.

A projected edition of the lexicographers never came to anything. But in this field too Bentley was of assistance to scholars in Holland and Germany, contributing to Küster's Suidas (1705) and Hemsterhuys's Pollux (1706); and his interest passed on to his successors. For nearly all the English and Dutch scholars of the eighteenth century are known as students and critics of the lexicographers. Toup's chief work was on Suidas, Porson edited Photius, and finally Gaisford produced editions of Suidas, the *Etymologicum Magnum*, and the Paroemiographi—all without the benefit of the 'team' work of the nineteenth and twentieth centuries.

The completely preserved texts of which Bentley published new editions were Horace (1711, often reprinted up to 1869), Terence, with a hastily added recension of Phaedrus and Publilius Syrus (1726), and Manilius (1739). His preface and notes to Horace contain the most famous saying about his 'method', which has been quoted again and again for more than two centuries, and has suffered the traditional fate

---

[1] *Ep. ad Millium* (above, p. 145 n. 5) pp. 351 ff.   [2] See Call. II pp. xlv f.

[3] See for instance Call. fr. 64.8, 13, 14; 178.33; 191.10.

[4] A characteristic and instructive example is Bentley's treatment of a Callimachean line (fr. 21.3) in the scholia to Lycophron. He recognized a genuine crux, but tried to cure it in the wrong place. The results can easily be traced in O. Schneider's note on his Call. fr. 206, as Schneider faithfully recorded Bentley's own words along with the attempts of later scholars. I hope I have correctly restored the hexameter of Callimachus by writing ἀνιήσουσα for the manuscript readings ἀνήσουσα or ἀνίσχουσα.

[5] See above, p. 109.   [6] On Leclerc see above, p. 137.

of such quotations by becoming half-quotation.¹ It is usual to quote (from the notes to Hor. *c.* III 27.15) only the words 'nobis et ratio et res ipsa centum codicibus potiores sunt',² and to leave out 'praesertim accedente Vaticani veteris suffragio'. Bentley, who was the first to recognize the outstanding value of the codex Blandinius vetustissimus for the text of Horace, was well aware of the need to consult the manuscripts before exercising criticism.

But he had hardly any doubt of the correctness of the text restored by his criticism; on the contrary he had complete confidence in his own 'divination'. On his conjecture of the rare word *vepris*³ for the manuscript reading *veris* at *c.* I 23.5, he commented: 'nihil profecto hac coniectura certius est; suoque ipsa lumine aeque se probat, ac si ex centum scriptis codicibus proferretur.' In the preface to his Horace there is a sort of climax in his admonitions to the scholar starting 'noli . . . librarios solos venerari; sed per te sapere aude'; going on to 'sola ratio, peracre iudicium, critices palaestra', and culminating with the need for divination, μαντική, which cannot be acquired by labour and long life, but must be innate. Confidence in his own divination led Bentley to the belief that he knew what the poet ought to have written. In Bentley's view Horace as a classic poet could not have written anything inconsistent with the harmonious measures of classical poetry. Here we see the significance for literary criticism of the passage of the Boyle lectures quoted above.⁴ The true critic must recognize the errors of transmission and restore the original harmony.

Bentley is said to have made more than 700 changes in the text of Horace following his intuitive method.⁵ It is easy for us to see its weakness; it was not controlled by knowledge either of the historical and individual style of the writer or of the history of the text. But historical understanding is, as we shall see, an acquisition of a later age. Bentley, in spite of his numerous and sometimes violent mistakes, did more than anybody before him to raise the standard of critical sense, and it is always profitable to follow up his suggestions on language,

¹ Cf. above, p. 73 n. 8 (Erasmus) and p. 116 n. 3 (Scaliger).

² One of Erasmus's correspondents, the Spanish theologian J. G. Sepulveda, had used a similar phrase about 'ratio' nearly two centuries earlier in speaking about interpretation and translation (Erasm. *Ep.* 2938.27 f., 23 May 1534, quoted above, p. 95, with other references to Sepulveda, see esp. *Ep.* 2905.16 ff. on Paul's Gal. 4 : 25). It is interesting to note in how many connections we find ourselves referring back to Erasmus and his circle.

³ It does not matter for our purpose that he had been anticipated by Gogavius.

⁴ Above, pp. 146 f.

⁵ R. Shackleton Bailey, 'Bentley and Horace', *Proceedings of the Leeds philosophical and literary society* 10 (1962) Part III, pp. 105–15.

grammar, style, and metre. As a conjectural critic he is without parallel in the history of classical scholarship.

It was mainly for metrical reasons that he changed almost a thousand readings in his edition of Terence (1726). But we must remember that the sixteenth and seventeenth centuries had recognized only two metres in Terence, the iambic and trochaic. Bentley, in the 'De metris Terentianis σχεδίασμα' which precedes his text, was the first to throw clear light on the metrical system, not only of Terence, but of the Latin dramatists in general. His discovery had revolutionary consequences for the constitution of the text; and we must not be surprised if it tempted him into some misleading statements. In the 'Schediasma' Bentley also inevitably came up against the tantalizing problem of the ictus in Latin and even in Greek verse.[1] Despite the efforts of modern scholars to reach a plausible solution, confusion and mistakes seem to have increased, and if we want to see things clearly, we had better go back to Bentley's 'Schediasma' of 1726. It remains a testimony of his ability to give a clear and satisfactory account of the most difficult subjects.

His editions also profited from that familiarity with Greek grammarians which had been obvious even in his earliest publications. The ancient Latin commentary of Donatus on Terence had been known in the fourteenth century to Salutati,[2] but Bentley was the first to make proper use of it for the text of the comedies.[3] Bentley's last edition, the Manilius[4] of 1739,[5] showed a new critical tendency; no fewer than 170 verses were rejected as interpolations.

His series of complete editions was accompanied by some minor but weighty contributions to the text of Suetonius'[6] *Caesares*, Cicero (in the Appendix to Davies's *Tusculan Disputations*, 1729), Nicander (for Dr. Mead in 1722), and various inscriptions, especially that on the Naxian colossus at Delos.[7]

---

[1] Cf. E. Kapp, 'Bentley's Schediasma "De metris Terentianis" [reprinted by F. Reiske in his edition of Plaut. *Rud.* (1826) pp. 77 ff.] and the modern doctrine of ictus in classical verse', *Mnemosyne* Series III 9 (1941) 187–94 = E.K., *Ausgewählte Schriften* (1968) pp. 311–17.

[2] See above, pp. 25 ff.

[3] Cf. G. Jachmann, *RE* v A 1 (1934) 598 ff. *passim*. K. Dziatzko, *Neue Jahrbücher*, Suppl. 10 (1878/9) 662 ff., 675 ff. referred to the Codex Bodleianus canonicus lat. 95 (s. xv) and to Bentley's conjectures.

[4] Cf. above, p. 117 (Scaliger).

[5] Between the Terence and the Manilius he had produced his regrettable and much criticized edition of Milton's *Paradise lost* (1732) which does not concern us here in a history of classical scholarship.

[6] M. Ihm, 'R. Bentley's Suetonkritik', *Sitz. Ber. Preuß. Akad. d. Wiss.* Jg. 1901, I 677–95.

[7] 'The image of the Delian Apollo and Apolline ethics', *Journal of the Warburg and Courtauld Institutes* 25 (1952) 20 ff. = *Ausgewählte Schriften* (1960) pp. 55 ff. with plates; on Bentley's discovery, p. 23. The inscription, probably seventh century B.C., on the eastern side of the

In the last decades of his life Bentley undertook two ambitious projects that had an immense effect on the future of scholarship, namely editions of the New Testament and of Homer. His friend John Mill[1] had published an edition of the New Testament in 1707. The text was of course the so-called *textus receptus*,[2] but in the footnotes Mill had recorded many more variant readings than any of his predecessors, collected by the labour of thirty years. Their number—about 30,000—was alarming; not only had it become impossible to find one's way through this labyrinth, but deists and free-thinkers, such as Antony Collins in his *Discourse on Freethinking*, derived much support from these 30,000 variants. Bentley in his reply to the *Discourse*[3] (1713) refuted Collins's arguments, but insisted, like a true inheritor of the Erasmian tradition, on the necessity of critical studies in their application to Scripture. After a visit from the young Swiss theologian J. J. Wetstein[4] Bentley's plan was definitely formed: he would not present once again the received text with a farrago of readings from manuscripts of all ages, but would try to restore the oldest knowable text. This was in his opinion the text of the fourth century A.D. at the time of the Council of Nicaea. He proposed to restrict himself to the oldest Greek manuscripts, supplemented by the oldest manuscripts of the Vulgate, of the ancient Oriental versions, and of the earliest quotations in the writings of the Church Fathers. The edition was to become, as Bentley said, 'a Charter, a Magna Carta to the whole Christian church'. He collected material from manuscripts for more than twenty years, zealously assisted, among other fellow labourers, by the French Benedictines.[5] Although personal difficulties, as well as the complexity of the problems, prevented Bentley from completing and publishing his edition,[6] his project anticipated by a whole century the work of Lachmann and others.

In 1726, while still preparing the New Testament, Bentley was meditating an edition of Homer; in 1732 he apparently began to

---

base of the so-called colossus of the Naxians of which substantial fragments still greet the modern visitor to the Temenos of Apollo at Delos, was first copied by a French traveller, the botanist Tournefort, in the seventeenth century. Bentley recognized it to be verse, rather bad verse, but still an iambic trimeter, proudly claiming: 'I am of the same stone, statue and base.' The foremost epigraphists of the time, including Montfaucon (see above, pp. 131 f.), had been unable to decipher and explain the line (Bentley, *Correspondence* II [1842] 589 f.; cf. Monk II 160 f.). So Bentley's success won him great admiration.
  [1] See above, p. 145 and note; on Mill's edition of the New Testament see A. Fox (above p. 145 n. 5), pp. 36 ff. and B. M. Metzger, *The Text of the New Testament* ([2]1968) 107 f.
  [2] See above, p. 108.   [3] See above, p. 148.
  [4] Cf. Monk I 397; II 120 f.   [5] See above, pp. 131 ff.
  [6] From the material left to his nephew, and now in Trinity College library, extracts were printed by A. A. Ellis, *Bentlei critica sacra* (1862).

prepare it, and we know that he was at work on it in 1734; but nothing was published. We have only his marginal notes to the edition of Homer in the *Poetae Graeci* of Henricus Stephanus, a quarto manuscript book with notes on *Iliad* I–VI, and some notes dealing specifically with his most famous discovery, the digamma. The first hint of this discovery, however, had occurred already in a copy of Collins's *Discourse on Freethinking* (1713),[1] in which he had written the passing remark: 'δίγαμμα aeolicum οἶνος Ϝοῖνος vinum'. In his quotations from Homer in his edition of Milton's *Paradise Lost* (1732) he printed it in the form of a capital Latin F. The digamma is nowhere to be found in the manuscript tradition of the Homeric text; but from the grammarians and inscriptions Bentley knew that this letter, denoting the sound of Latin *v*, had existed. For linguistic and metrical reasons he reintroduced it, and the importance of this discovery is enormous; for the first time a step had been made beyond the text as it had been fixed by the Alexandrian grammarians and their followers in later ancient and medieval times.

Classical scholars were reluctant to accept what seemed to them a figment of prehistoric speculation. Richard Dawes, who criticized Bentley in part four of his *Miscellanea critica* (1745, and often reprinted), was a prominent opponent in England. In Germany a great admirer of Bentley,[2] F. A. Wolf, who was the first to try to restore the Alexandrian text, was not able to grasp the significance of the discovery and dismissed it as a 'senile ludibrium Bentleiani ingenii'.[3] Slowly, however, understanding[4] and at last acceptance came with the development of linguistic studies.[5]

Shortly before the period, in the years around 1726, when Bentley was considering a new edition of Homer, Alexander Pope (1688–1744) had translated the *Iliad* (1720) and the *Odyssey* (1725/6). Bentley's judgement about this translation is well known:[6] 'a very pretty poem . . . , but he must not call it Homer'. How carefully Pope had worked in translating Homer is attested by his extensive notes which were published with the first edition of the translation and several times

---

[1] See above, p. 148; cf. Monk II 363.   [2] See above, p. 148 n. 8.

[3] F. A. Wolf, *Kleine Schriften* II (1869) 1070. 'Spielwerke von unmäßiger Willkürlichkeit', ibid.

[4] An important step in that direction was made by J. W. Donaldson, *The New Cratylus* (1839) p. 118 and appendix pp. 138 ff. 'Extracts from Bentley's MS on the Digamma'.

[5] See E. Schwyzer, 'Griechische Grammatik', *Handbuch der Altertumswissenschaft* II. Abt. 1. Teil, 1. vol. (1939) 222 ff.; the best introduction to the whole problem, as it seems to me. G. Finsler, *Homer in der Neuzeit* (1912) pp. 309 ff. gives a sensible short survey.

[6] Monk II² (1833) 372.

reprinted. There seems to be no evidence that Bentley also read the learned 'Observations',[1] a continuous bulky commentary which he would hardly have liked.

It is interesting to note what Bentley thought of Homer as a poet. In his remarks on Collins's *Discourse on Freethinking*, where he first mentioned the digamma, Bentley exclaimed against the pompous phrase of Collins that Homer 'designed his poems for eternity to please and instruct mankind'.

Take my word for it, poor Homer had never such aspiring thoughts. He wrote a sequel of songs and rhapsodies, to be sung by himself for small earnings and good cheer, at festivals and other days of merriment. The Iliad he made for the men and the Odyssey for the other sex. These loose songs were not collected in the form of an epic poem till Pisistratus' time, above 500 years after.

The last sentence repeats the common tradition of later antiquity about the collection of the Homeric poems by Pisistratus;[2] the view expressed about the quality of Homeric poetry may be found at the same time in Leibniz, and especially in those French books that took the modern side in the 'Querelle des anciens et modernes'.

The key to Bentley's judgement on Homeric poetry can be found by putting it beside the judgement he had pronounced on the *Aeneid* in his Boyle lectures. It is quite obvious that the Greek epics lacked the 'good sense and just measures' he had appreciated in Virgil's classic poem. In this respect Bentley does not appear to be in advance of his age; for the view he held belonged to the seventeenth century. On the other hand the discovery of the digamma and its position in the Homeric language pointed firmly into the future.

[1] H. J. Zimmermann, *Alexander Popes Noten zu Homer* (1966), was the first to use Pope's autograph.
[2] *History* [1] 6 f.

XII

BENTLEY'S CONTEMPORARIES AND SUCCESSORS

THERE is in England a long and continuous history of classical education, of humanistic ideas, and of the classical tradition in literature; but the chapter formed by the history of classical *scholarship* is relatively short. There is not, as in other countries, a line of scholars of approximately equal merits, but just *one* figure incomparably greater than all the others. The effect made by his books and by his personality was such that he had understandably been extolled as the founder of modern classical scholarship.[1] This, however, seems to be an over-simplification of the historical process. For we have come across a number of scholars who made their contributions to significant changes in classical scholarship which resulted in its new form as a whole. When we look back to Valla, Erasmus, and Scaliger, we can see Bentley's work as the culmination of a historical process; yet his greatness was felt immediately. The impression he made is evident in the letters of his leading contemporaries,[2] and it was by no means confined to his own country, but spread abroad, principally to Holland and Germany. Bentley had no outstanding pupils of his own, but every classical student felt the inspiring influence of his critical sense. Textual criticism became the chief interest in England as literary criticism did in Germany. English textual criticism was mainly applied to the Attic poets to whose study Bentley had pointed the way in the *Epistula* and in the *Dissertation*. Although he left nothing like a 'school' behind him, Cambridge[3] remained after his death a centre for classical scholars, of whom Richard Porson (1759–1808) was the greatest.

Porson's earliest enterprise, an edition of Aeschylus, was frustrated by a series of accidents. But in his textual notes, written from 1783 onwards

---

[1] *Proceedings of the Classical Association* 59 (1962) 25 ff.

[2] C. B. Hunt, 'Contemporary References to the Work of R. Bentley', *Bodleian Library Record* 7 (1963) 91 ff. (from the forties to the eighties of the eighteenth century).

[3] See Sandys's (he was a Cambridge man himself) illuminating pages II 422 f. on Cambridge.

and made known by his friends much later, there were astonishing feats of emendation, comparable in quality to those of Jean Dorat[1] (who also had been unable to publish his notes himself). Bentley had led the way in the scholarly treatment of Greek and Latin metre,[2] and Porson was the first to make a further substantial advance; we may say that his claim to immortality is based above all on the rule called 'Porson's Law': that no word may end after a long *anceps* in the last iambus of the tragic iambic trimeter.[3] In a sense he followed in Bentley's traces also in his *Letters to Travis* on the Comma Ioanneum (1 John 5:7 and 8); Erasmus had rejected the Comma as an interpolation, and so had Bentley.[4] But as Travis tried to prove its genuineness, Porson wrote in twelve months of 1788–9 a book—his only one—of 400 pages in which he proved the spuriousness of the Comma Ioanneum in all its details, with an enormously learned apparatus and acute textual observations, applauded by Gibbon.

The young Porson, although coming from a poor family, was extremely fortunate at the start of his career: wealthy patrons helped him to go to Eton and then in 1778 to Trinity College, Cambridge; in 1782 he became a fellow of his college and in 1792 Regius Professor of Greek. He never did any lecturing or teaching;[5] but his reading and his memory were unlimited. Living in London, he became a lion in society as a brilliant talker, and occasional flashes of genius revealed the born scholar in him. After a time of physical and mental decline he ended as a helpless and hopeless drunkard. It is only from the sombre story of his life[6] that one can try to understand the grievous barrenness of his later years; 'tragic' is hardly too strong a word for such a waste of talent. Whatever he succeeded in finishing during his best years consisted of accurate observations made with caution and patience, not improvisations or hasty suggestions. For many he was the ideal of exact verbal scholarship. Most of his critical notes, starting with the *Adversaria* (1812), were published after his early death by his devoted friends. What Bentley had regarded as the decisive quality of a true scholar,

---

[1] See above, p. 104. [2] See above, pp. 149 f.

[3] Eur. *Hec.* ed. Porson (London 1797) on l. 347; Eur. *Phoe.* (1799) on l. 1464 and later; see P. Maas, *Greek Metre*, translated by H. Lloyd-Jones (1962) pp. 34 f. Besides *Hecuba* and *Phoenissae* the famous quartet of Euripidean editions by Porson included *Orestes* (1798) and *Medea* (1801).

[4] See above, p. 148.

[5] Cf. below, p. 181 Bekker.

[6] J. S. Watson, *The Life of Richard Porson* (1861); M. L. Clarke, *R. Porson, A Biographical Essay* (1937); Denys Page, 'Richard Porson', *Proceedings of the British Academy* 45 (1959) 221–36, so far the most competent appreciation.

the power of divination,[1] was possessed by Porson in the highest degree.[2] No wonder therefore that he dominated Greek studies for several decades.

Monk,[3] C. J. Blomfield (1786–1857), and Elmsley continued Porson's work on the dramatists: Elmsley (1773–1825) in Oxford. Indeed the text of the Greek dramatic poets was constantly being worked on in England before and after Bentley.[4] Already Thomas Stanley[5] (1625–78) had started to apply serious criticism to Aeschylus, Benjamin Heath (1704–66) had tried to clear up the problem of the manuscript tradition of Sophocles, and Josuah Barnes (1654–1712) had published his complete one-volume edition of Euripides (1694). This was followed by Samuel Musgrave's (1732–81) *Exercitationes in Euripidem* (1762), with an appendix by Tyrwhitt, whose high reputation extended beyond his own country. Dobree (1782–1825) worked on the Attic orators. Jeremiah Markland (1693–1776) was the only British classical scholar after Bentley who was equally eminent in both Greek and Latin, and in contrast to him a modest and amiable personality. Richard Dawes[6] (1709–66), although he did not finish any book, had his various articles published in his *Miscellanea critica* (1745)[7] dealing mainly with problems of Greek grammar.

Looking back at this line of creative British scholars during the eighteenth century we may be inclined to accept Bywater's[8] admission that their writings, especially those of the Porson school, reveal a certain insularity and narrowness. But we must not forget that at the same time Robert Wood (1717–71), though travelling as a politician, visited the Near East, especially the Troad, with Homer in his pocket and produced in his 'Essay on the original genius and writings of Homer'[9] an archaeological and historical book that was widely read in the original and in translations and of decisive influence on classical studies. Even before Wood's work became known the 'Society of Dilettanti' (founded probably in 1732) had started in 1762 to publish a magnificent series of folio volumes with drawings of works of Greek art, which were of

[1] See above, p. 154.   [2] Aesch. *Ag.* 1391 f., see E. Fraenkel, Aesch. *Ag.* 146 f.
[3] See above, p. 148 n. 8.
[4] On Bentley himself see above, pp. 149 ff., 152.
[5] Cf. above, p. 152, and *History* [I] 134.6, and the supplement in the German translation (1970) Exkurs II p. 339.
[6] See above, p. 157.
[7] New edition 1781, several times reprinted.
[8] *Four Centuries of Greek Learning in England*, Inaugural lecture of 1894, published in 1919. M. L. Clarke, *Greek Studies in England 1700–1830* (1945) pp. 48 ff.
[9] Privately printed in 1767, published posthumously in 1775 and later.

fundamental importance for English classicism.[1] The greatest sensation, of course, was and will always remain the original sculptures from the Acropolis, called the Elgin marbles, in the British Museum, brought to England by Lord Elgin.[2]

Towards the end of the century, in the years 1776–88, a historian, the greatest of the century, as many believe, produced one of the most impressive books ever written on the ancient world—Edward Gibbon (1737–94), *The History of the Decline and Fall of the Roman Empire* in six quarto volumes.[3] The range of his reading in Latin and Greek widened and deepened during his years in Lausanne; he also admired Porson and defended him against attacks. Porson in turn wrote a splendid review of Gibbon's work.[4] For Bentley[5] Gibbon found the most appropriate epithet, 'tremendous'. It is clear, too, from his great work that his qualifications for writing it included an acquaintance with the history of scholarship.

Holland, with which Thomas Tyrwhitt (1730–86) and Samuel Musgrave (1732–80) had special personal and scholarly relations, and where we found Johann Georg Graevius[6] (1632–1703) editing Callimachus with the help of Bentley's collection of the fragments (1697), remained on the whole conservative. Although he was acquainted with Bentley, Pieter Burman (1668–1741), whose *Sylloge Epistularum a viris illustribus scriptarum* (1724) in five quarto volumes is the most valuable source for the history of Dutch scholarship, continued to represent the traditional leaning towards polymathy and the manufacture of bulky 'Variorum editiones'; and the same is true of his nephew Pieter (1714–78). Yet there was a notable difference between the work of some Dutch scholars in the eighteenth century and that typical of the seventeenth, as the rigidity of the earlier encyclopedism had been relaxed by the infiltration of the new critical spirit inspired by Bentley. In Rotterdam the Westphalian Ludolf Küster (1670–1705) wrote his Suidas with Bentley's assistance (it was published in three folio volumes in 1705 and continued by Toup in the late eighteenth century)

---

[1] Lionel Cust, *History of the Society of the Dilettanti*, ed. Sidney Colvin (1914). Cf. B. H. Stern, *The Rise of Romantic Hellenism in English Literature, 1732–1786* (1940, reprinted 1969).

[2] William Saint Clair, *Lord Elgin and the Marbles* (London 1967).

[3] The standard edition is that of J. B. Bury, 7 vols. (1896–1900). The *Letters* ed. J. E. Norton, 3 vols. (1956). *Autobiographies*, ed. John M. Murray (1896). Jacob Bernays unfortunately never finished his 'essay' on Gibbon, but substantial fragments are preserved in *Gesammelte Abhandlungen* II (1885) 206–54. David P. Jordan, *Gibbon and his Roman Empire* (1971).

[4] Reprinted in Watson's biography of Porson (above p. 160 n. 6) pp. 85 f.

[5] See above, p. 143, and Michael Joyce, *E. Gibbon* (1953) p. 65 'the tremendous Bentley' (without references).

[6] See above, p. 152 and below, p. 163 n. 1 'Schola Hemsterhusiana' (1940) pp. 18–27.

and prepared Hesychius for the printer. Still more the work of the triad of great scholars in Holland, Tiberius Hemsterhuys (1685–1766),[1] L. C. Valckenaer (1715–85), and David Ruhnken (1723–98),[2] clearly shows how the study of the Greek lexicographers initiated by Bentley enabled his successors to produce better editions of the ancient Greek lexica and to recognize the idiomatic peculiarities of Greek, especially Attic, poets and writers. This line of scholarship descended through Wyttenbach (who was by birth a Swiss) to Cobet (1813–89) in the nineteenth century, and still continues in the twentieth. In the field of Roman history and literature Perizonius (Jacob Voorbrock), in his *Animadversiones historicae* (1685), showed a critical spirit in advance of his age.

In Germany the influence of Bentley could not be immediately felt and did not become effective until the very end of the eighteenth century in F. A. Wolf and G. Hermann. Classical studies had entered a period of inactivity, and there was a little feeling for the ancient world, although a very few learned men were at work at Vienna and Hamburg, at Leipzig and Göttingen. But it was not from these learned men that the resurgence of classical studies came nor yet from outside Germany.[3] It was the consequence of a completely new approach to antiquity by one man of genius, comparable only to the fresh start made by Petrarch in the fourteenth century.

[1] J. G. Gerretzen, 'Schola Hemsterhusiana', *Studia Graeca Noviomagensia* 1 (1940) 77–156.

[2] On Ruhnken as inventor of the term 'canon' for selective lists of Greek authors in his *Historia critica oratorum Graecorum* (1768) see *History* [1] 207. E. Hulshoff, *Studia Ruhnkeniana* (1953) pp. 142 ff. It was Ruhnken as 'princeps criticorum' to whom F. A. Wolf dedicated his *Prolegomena ad Homerum* (1795).

[3] I should like to refer to the very instructive paper of Harald Keller on English classicism and its influence on Continental architecture: 'Goethe, Palladio und England', *SB der Bayer. Akad.*, Phil.-hist. Kl., Jg. 1971, Heft 6.

PART FOUR

GERMAN NEOHELLENISM

XIII

WINCKELMANN, THE INITIATOR OF NEOHELLENISM

THE study of classics has never been extinct in Germany, but it lived
on in a quiet and modest way throughout the seventeenth century. If it
is correct to call that century the age of the scientific revolution,[1] the
eighteenth century might be described as the age of humanistic
revolution.[2] The masterpieces of classical literature once again
produced a miraculous quickening of the spirit, as in the time of
Petrarch;[3] now, however, the source of inspiration was not Virgil, not
Cicero, neither Roman sweetness nor sonority, but Homer, Sophocles,
Herodotus, and Plato, 'the noble simplicity and serene greatness' of
the Greeks, in Winckelmann's own famous phrase.[4]

Johann Joachim Winckelmann (1717–68) was born in Stendal,[5]
west of Berlin, the son of a poor cobbler. During his youth he encountered various great difficulties, which only his boundless enthusiasm
and energy enabled him to surmount. It was very hard for the young
student to obtain Greek texts; he had to exist mainly on printed
anthologies and on excerpts[6] written by his own hand. For it was only
towards the end of the century that F. A. Wolf and his pupils began
to prepare handy Greek texts for German publishers on a large scale.

Places of learning were distant, and he had to travel by foot; the
nearest was Berlin, not yet of major importance in about 1730. But the
head of the oldest high school there, Christian Tobias Damm (1698–
1778), loved the Greek language above all and was even described as

---

[1] H. Butterfield, 'The history of sciences and the study of history', *Harvard Literary Bulletin* 13
(1959) 346.

[2] See below, p. 169.   [3] See above, pp. 4 ff.

[4] Cf. below, p. 169. It is usually assumed that it was coined by Oeser (see Justi 1² 322,
below, p. 168 n. 5), but perhaps its ultimate source was French. On Winckelmann's French
sources see A. Buck, *Die humanistische Tradition in der Romania* (1968) *passim*, and the review of
Buck's book by Stackelberg, *Gnomon* 42 (1970) 424 ff.

[5] Henri Beyle called himself Stendhal [*sic*] in honour of Winckelmann; on the pseudonym
see G. von Wilpert, *Lexikon der Weltliteratur* 1 (1963) 1270. On the precarious situation of
scholarship in Germany see above, pp. 141 f.

[6] W. Schadewaldt, 'Winckelmann als Exzerptor' in *Hellas und Hesperien* (1960) 637–57.

"Ὁμηρικώτατος': he published an etymological Homeric dictionary and a translatioᵑ of the *Iliad* and the *Odyssey* into German prose (1767), which found many readers, and he even ventured to prophesy: 'videor iam saeculum renascentis apud nos Graecitatis cernere animo.'[1] Winckelmann had only a small selection of books when he settled down in 1742 as 'Konrektor' in Seehausen north of Berlin, but he read them again and again, not only the Greek and Latin classics, but also the great scholars of the French Renaissance, J. J. Scaliger, his friend de Thou and his pupil Grotius.[2]

J. M. Gesner (1691–1761), who taught classics at Göttingen from 1734 onwards has been described as a 'precursor' of Winckelmann.[3] This is incorrect. For Winckelmann continued his studies not at Göttingen, but at Halle and Jena, and there were no personal or scholarly relations between the two men. Yet students who had attended Gesner's classes in Leipzig, and especially in Göttingen, found it easier to comprehend and to accept the ideas of Neohellenism. Winckelmann had indeed intended to come to Gesner's seminar in Göttingen. But instead he had the good luck to join the staff of the library of count Heinrich von Bünau[4] near Dresden, probably the greatest private library in Germany at that time and of priceless value to a voracious reader like Winckelmann. From poetry, history, and scholarship he moved on to the study of Greek art in Dresden (when he was staying near the City in Bünau's service) and finally in Rome. His life, more and more dedicated to the intensive study of Greek art, was full of strange chances. Not being a Catholic, he had at first no access to the main art collections in Rome; then, despite his complete and deliberate paganism, he was received into the Roman Catholic Church, and many unexpected honours fell to his lot, such as the post of librarian in the Vatican and the presidency of antiquities. Equally unexpected was his tragic end: he was horribly murdered by an Italian cook at Trieste while on his way from Vienna to Rome in 1768.

Winckelmann did not publish anything until he had finished his various studies and was on his way to Rome in 1755.[5] His first publica-

---

[1] In the *Program* 1752, quoted by Justi (below, n. 5 [on p. 169]) I² (1898) 34.2; 'video' in the *Program*, corr. J. K. Cordy.

[2] See above, p. 119.

[3] As Sandys III 7 says.

[4] See Justi I² 181 ff. and the indexes of the four volumes of Winckelmann's letters (below n. 6).

[5] *Sämtliche Werke*, Einzige vollständige Ausgabe hg. von J. Eiselein, Donauöschingen [*sic*] 1825–9, 12 vols. (reprinted 1965). *Werke*, einzige rechtmäßige Originalausgabe, 2 vols. (Stuttgart 1847); this is the edition I have used. *Briefe*, hg. von W. Rehm in Verbindung mit

tion dealt with the central problem of μίμησις: *Gedanken über die Nachahmung der griechischen Werke in der Malerei und Bildhauerkunst.*[1] The complete Winckelmann, investigator and seer, and above all master of language, can be found in this small pamphlet of 1755. From 1742 onwards we can read his letters.[2] Winckelmann grew up at the beginning of the age of German classical poetry; he did not write poetry himself, but he did write the prose of a poet. The essay on 'Nachahmung' contains characteristic and beautiful phrases which have remained alive: 'Der einzige Weg für uns, groß, ja wenn es möglich ist, unsterblich zu werden, ist die Nachahmung der Alten.'[3] On the ancients themselves he coined the simple and monumental sentence:[4] 'Die edle Einfalt und stille Größe der griechischen Statuen ist zugleich das wahre Kennzeichen der griechischen Schriften aus den besten Zeiten.' It may be that Winckelmann owed more than we can prove now to the conversations which he had with his great friend, the artist Friedrich Oeser,[5] near and in Dresden, before he travelled to Rome. In several cases the original source may have been French, as he was thoroughly familiar with the French language and literature.[6]

Winckelmann's first publication was hailed by Herder in his 'Preisschrift' (Cassel 1778) as his 'vielleicht seelenreichstes Buch . . . und duftreichste Jugendblüthe'. But nobody paid a higher tribute to Winckelmann than Goethe when he gave to the memorial essays[7] that he published in 1805 the title *Winckelmann und sein Jahrhundert.* The eighteenth century was in many respect a great century; for Goethe it was the century of Winckelmann.

Goethe achieved in his poetry what Winckelmann had hoped that a

H. Diepolder, Berlin 1952–7, 4 vols.; in IV 369 ff. 'Urkunden und Zeugnisse zu Winckelmann's Lebensgeschichte'. Carl Justi, *Winckelmann und seine Zeitgenossen*, 3 vols., 1866–72, 2. Aufl. 1898 which I used, 5. Aufl. von W. Rehm 1954 with a preface about Justi and the earlier editions. A selected bibliography up to the present time in W. Leppmann, *Winckelmann* (New York 1970); see especially W. Rehm, *Griechentum und Goethezeit* (1936, 4th ed. 1968) and A. Buck, *Die humanistische Tradition in der Romania* (1968) pp. 122 ff. and *passim*, and cf. the review of J. von Stackelberg (see above, p. 167 n. 4). Cf. A. H. Borbein in his review of U. Hausmann, 'Allgemeine Grundlagen der Archäologie' in *Handbuch der Archäologie*, 1969, *Gnomon* 44 (1972), especially pp. 287 ff. on Winckelmann. (A curiosity, not mentioned in the bibliography, is E. M. Butler, *The Tyranny of Greece over Germany* (1935) pp. 9–48, an anti-Winckelmann polemic concluding 'it is enough to make the merciful regret that Winckelmann was ever born.')

[1] Reprinted together with 'Sendschreiben' and 'Erläuterung', *Werke* (Stuttgart 1847) II 1–57, cf. Justi I² 351–403, see especially p. 394.

[2] See above, p. 168 n. 5; on the style see especially W. Rehm's introduction to the letters, vol. I (1952).

[3] *Werke* II p. 6, § 6.   [4] *Werke* II p. 13, § 88; see above, p. 167 n. 4.

[5] See Justi I² 316 ff.   [6] See above, p. 167 n. 4.

[7] Goethe, *Sämtliche Werke*, Jubiläumsausgabe, XXXIV 1–48.

great artist would achieve in new classical works of art. In contrast with the modern cult of originality Winckelmann assigned a high place to imitation in Greek art and literature, as it kept alive the best of the tradition.

Only in emulation of earlier masterpieces could new ones be created. Roman culture now appeared to be no more than an approach to that of Greece. A break was made with the Latin tradition of humanism and an entirely new humanism, a true new Hellenism, grew up. Winckelmann was the initiator, Goethe the consummator, Wilhelm von Humboldt in his linguistic, historical, and educational writings the theorist. Finally Humboldt's ideas were given practical effect, when he became Prussian minister of education and founded the new university of Berlin and the new humanistic gymnasium.

Winckelmann's influence was not only humanistic, but at the same time also historical. He was the first to treat history organically as the record of the growth of the human race. His *History of Ancient Art*[1] (1764) embraced the development of the art of Egypt, of Phoenicia and Persia, of Etruria, and finally, of Greece and Rome. In the outstanding part of this work, in that on Greece, he took up a suggestion made by J. J. Scaliger[2] that there were four ages of Greek poetry, and distinguished four different styles of Greek art as developing in harmony with the national life as a whole.

Winckelmann's *History of Ancient Art* was received with the greatest enthusiasm all over Europe. This universal success would not have been possible if the earlier humanism of the Renaissance had not formed a supranational unity of the spirit which had saved Europe from complete cultural disruption and which still existed in Winckelmann's day. The most competent of its earliest readers was Lessing, whose *Laokoon* (1766) started a fruitful debate on the *History* in which it was answered by Herder's *Kritische Wälder* (1769).

After two years in Rome Winckelmann visited Naples for three months and went on from there to the excavations of Herculaneum and Pompeii, then to Paestum and Agrigentum. These visits were followed by a number of smaller publications: *Anmerkungen über die Baukunst der alten Tempel zu Girgenti in Sizilien* (1759), *Description des pierres gravées du feu baron de Stosch* (1760), *Anmerkungen über die Baukunst der Alten* and *Erinnerung über die Betrachtung der Werke der Kunst* (1761), *Von der*

---

[1] *Geschichte der Kunst des Altertums* (1764), vollständige Ausgabe von W. Senff (1964).
[2] See above, p. 119. I repeat that it was Winckelmann, not Niebuhr, who was the first to appreciate the unique greatness of Scaliger.

Fähigkeit der Empfindung des Schönen and *Von der Grazie in den Werken der Kunst* (1763). In the years 1762-3 he was mainly occupied with the completion of his history of ancient art, which he published in 1764, and to which he added in 1767 his *Anmerkungen zur Geschichte der Kunst.* Between these two books his *Versuch einer Allegorie der Kunst* came out in 1766. The crowning masterpiece of his years in Italy before his death was his *Monumenti antichi inediti* in two volumes (1767/8), which laid the foundation of 'monumental mythology'. From the monuments visible in southern Italy, he drew the attention to the still-invisible treasures below the soil of Olympia and in an evocative vision of the greatest discovery of the future, declared: 'Ich bin versichert daß . . . durch genaue Untersuchung dieses Bodens ein großes Licht aufgehen würde.'[1]

The new humanistic approach, first inspired by the study of Greek poetry,[2] and then applied to art, became fruitful again for literature in the writings of Lessing, Herder, and Friedrich Schlegel, and in the literatures of other European countries. In Germany, but nowhere else, there grew up a sort of evangelistic humanism which was both warmly espoused and bitterly attacked for several generations. It was a powerful movement, which, headed by Winckelmann, took its place beside the systems of the leading philosophers from Kant to Hegel; and it was this power that renewed classical scholarship in Germany.

A very gifted classical scholar inherited the leadership, Christian Gottlob Heyne (1729-1812), who like Herder devoted his prize essay[3] of 1778 to Winckelmann. Heyne was born in Chemnitz, studied in Leipzig, and became active first as a copyist in Dresden in the library of count Heinrich von Brühl where he made Winckelmann's acquaintance.[4] How deep and lasting the impression on Heyne was, his prize essay of 1778 shows. It was precisely the influence of Winckelmann[5] that distinguished the scholarship of Heyne and his friends and pupils from that of other contemporary scholars.

The scholar who had most in common with Heyne, and much also with Gesner was J. A. Ernesti (1707-81). Coming from Schulpforta, which in the eighteenth and nineteenth centuries produced so many

---

[1] Max Wegner, *Altertumskunde* (1951) p. 122. The German excavations began about a century after Winckelmann's *Monumenti* and were, after another century, renewed in our own day (by E. Kunze).

[2] See above, p. 167.   [3] See above, p. 169.   [4] *Briefe* IV 454.

[5] This is not sufficiently or not at all acknowledged in the monographs on Heyne. F. Leo, 'Heyne', *Festschrift z. Feier des hundertfünfzigjährigen Bestehens der kgl. Gesellschaft der Wiss. zu Göttingen* (Berlin 1901) pp. 153-234. F. Klingner, 'Christian Gottlob Heyne', *Studien zur griechischen und römischen Literatur* (1964) 701 ff.

distinguished scholars and schoolmasters, he made many useful editions of Greek and Latin texts. He was a member of a great family of scholars, of whom his nephew Christian Gottlieb Ernesti (1756–1802) deserves special mention, as it is his *Lexiçon technologicum* that we still use for the study of Greek and Latin rhetoric; no one has attempted to supersede it. People like the industrious and solid Ernestis occupied the chairs of the bigger universities, while incomparably greater scholars like Reiske never succeeded in doing so.

J. J. Reiske (1706–74) had no real contact either with Winckelmann himself or with the new movement as a whole, but remained a distinguished outsider.[1] Only a few of his best contemporaries, the Prussian King Friedrich II and Lessing, recognized his qualities, and they did not give him any help when he urgently needed it. After a period of serious financial difficulties he was finally appointed Rector of the Nicolai-Schule in Leipzig in 1758 and became able at last to make his invaluable contributions to the text of the Attic orators. In this he was continuing the work of Hieronymus Wolf,[2] whom he curiously resembled both in his character and in his unhappy life. He had to live partly on the fruits of his Arabic studies, and it is said that he was no less proficient as an Arabist than as a Greek scholar.[3] Lessing's plan to write a biography of Reiske in three volumes came to nothing, but at least we have the much shorter account of his life written by himself[4] and supplemented by his wife, who was deeply devoted and always helpful to him. Reiske's connections were with pre-Winckelmann scholarship, while all the other classical scholars of the time were in some degree followers of Winckelmann.

[1] There seems to be no evidence that he had read Winckelmann's publications with any care.
[2] See above, p. 140 and especially n. 1.
[3] See Johann Fück, *Die arabischen Studien in Europa* (Leipzig 1955) pp. 108 ff.
[4] *Reiskens von ihm selbst aufgesetzte Lebensbeschreibung* (1783). His letters were collected and edited by Richard Foerster, *Abhandlungen der philol.-hist. Klasse der k. Sächsischen Gesellschaft der Wissenschaften Bd. 16 (1897)* and 'Nachtrag', *ib. Bd. 34.4 (1917)*.

XIV

FRIEDRICH AUGUST WOLF

THE last and greatest of Winckelmann's followers was Friedrich August Wolf (1759–1824), and it was he who wrote Winckelmann's biography, as well as one of Bentley.[1] If Winckelmann's influence determined the direction which Wolf's work took, it was Bentley's critical spirit that prevented his scholarship from becoming too fanciful under the influence of humanistic enthusiasm.

Wolf was born at Hainrode, south of the Harz, in the same year as Porson and Schiller, 1759, and lived until 1824, the year before Elmsley and Dobree died. He went to school at Nordhausen, where his headmaster was a member of the family of the learned Fabricii, Johann Andreas (1696–1769), who had written an *Abriss einer allgemeinen Historie der Gelehrsamkeit* in three volumes (1752–4). Wolf matriculated in the university of Göttingen in 1777. Legends grew up about this matriculation,[2] but the simple fact seems to have been that, though he insisted on entering his name as 'studiosus philologiae';[3] he was not the first to do so and was not conscious of opening a new era of classical studies. He began to concentrate on Homer while attending Heyne's courses, and also on Plato, producing in 1782 an edition of the *Symposium*; in 1785 he started his own lectures on the *Iliad*. In the course of time he treated many subjects; as an impassioned lecturer he roused a new interest in ancient literature, and as a capable organizer he attracted students to his classes. But his greatest projects were for new texts of Homer and of Plato. This recalls Bentley's two great plans, for Homer and the New Testament, except that Wolf as a good pagan and classicist had no taste for the Greek of the Bible and chose Plato instead. He was not able to complete his projects, but his influence was such that Homeric and Platonic problems dominated classical studies for generations. What Wolf achieved, though only a part of his

---

[1] Of the two biographical essays one was published in the collection *Winckelmann und sein Jahrhundert* (1805), and the other on Bentley in Wolf's own German periodical *Lit. Anal.* 1 (1816) 1–89 = *Kleine Schriften* II (1869) 1030 ff.; cf. above, p. 169.

[2] Sandys III 52.

[3] Edward Schröder, 'philologiae studiosus', *Neue Jahrbücher für das klassische Altertum* 32 (1913) 168 ff.

programme, was a very important part: the small volume of his *Prolegomena ad Homerum* (1795), which won world-wide fame, contains the first[1] methodical and firmly based attempt at a history of an ancient text. It was based on the scholia on the *Iliad* just published by Villoison, who in 1788 had discovered in Venice the primary manuscript now called Codex Venetus A[2]. Wolf intended his history of the Homeric text to provide the basis for a judgement on the value of the manuscripts and for the constitution of the text that he wanted to publish. He came to the conclusion that it was impossible to reconstitute the text as it came from the hands of the author, but that we could try to restore the 'Alexandrian' text, that is, the text which the Alexandrian grammarians possessed in the third century B.C.

In tracing the history of the Homeric text between the age in which the *Iliad* and *Odyssey* were composed and that in which the poems were in the Alexandrian library and in the hands of the Alexandrian grammarians,[3] Wolf had to investigate the origin of the Homeric poems, and that involved the questions of their genuineness and of their unity. Wolf opened the eyes of the learned world to the fact that Homeric poetry holds a unique *historical* position and therefore cannot be studied with the same methods as Virgil and any later epics. No word is more stressed in the *Prolegomena* than the often repeated 'historia', mostly in the connection 'historiae et critices rationes'. The results of Wolf's researches through many years found their concentrated expression in his edition of 1795; in the preface to his *Homeri et Homeridarum Opera et Reliquiae* he stated firmly:[4] 'Tota quaestio nostra *historica et critica* est, non de *optabili* re, sed de re *facta* . . . Amandae sunt artes, at reverenda est historia' (Wolf's italics).

It is always a pleasure to read Wolf's unconventional, and yet clear and beautiful, Latin; but like Bentley[5] he did not confine himself to the traditional language of scholarship, but also used his native German,[6] as when he wrote about Winckelmann.[7] He never imitated Winckelmann's individual high style, but always remained true to his own very simple and impressive way of writing.

---

[1] On earlier attempts in the biblical field see above, p. 130.
[2] On this MS. see above, p. 48 and n. 2. Cf. also Ch. Joret, *D'Ansse de Villoison* (1910).
[3] See *History* [1] 105 ff.   [4] p. xxvi in the edition of 1804 (available to me).   [5] See above, p. 151.
[6] His lectures in German about *Encyclopaedia* (see below, p. 175), often repeated, were printed in 1831 and on p. 474 of vol. 1 of Wolf's *Vorlesungen über die Altertumswissenschaft* a passage on Cuiacius finished with the characteristic sentence: 'Er hatte eine erstaunlich liederliche Tochter', no doubt a casual remark in a lecture, of which unfortunately we have no details.
[7] See above, p. 173.

Wolf tried to give sober and exact proofs for the conclusion that there was not one single poet who wrote our *Iliad* and *Odyssey*, but that we owe these poems to a series of rhapsodes. He was not content to express general opinions and suggestions about the distinction between natural and artistic poetry, but endeavoured to prove his position by page after page of sound arguments. In the course of time much of his argumentation had to be retracted, and much was refuted. The abiding value of his work was in its spirit of critical and historical inquiry in which the essential connection of criticism and history was made.

The *Prolegomena* produced a sensation which spread far beyond the learned world, because it came out just at the right moment. The notion of original, popular, natural, naive poetry was on everybody's lips, and Homer was the poet most under discussion. Robert Wood's essay on the original genius of Homer[1] had had an enthusiastic reception, especially in Germany by Goethe and others; Herder had generalized and popularized the ideas in Macpherson's *Ossian* and in Percy's *Relics of Ancient English Poetry* and applied them to the Greek epics. On the other hand there were many who declared the *Prolegomena* a literary 'impiety'. But the most significant response was the fully sympathetic review written shortly after publication by Friedrich Schlegel, the most perceptive critic in the group of early German romanticists. He called the *Prolegomena* a work of more than 'Lessingschen Scharfsinns' and applied its principle to literary history in general.

The influence of the *Prolegomena* on the development of classical scholarship was incomparable. The so-called 'Homeric question' at once became one of the central problems and remained so until our own day. Although Wolf did not publish a critical analysis of the *Iliad*, but only some hints in that direction, he provided the impulse for the employment of the analytic method by generations of scholars in the epic field as well as in other provinces. It would be unjust to make him personally responsible for all the disastrous consequences of its application to unsuitable subjects.

All Wolf's achievements in particular areas of classical studies were subordinated to his general conception of these studies, for which he invented the comprehensive term 'Altertumswissenschaft'. Gesner[2] had entitled his introductory lecture *Isagoge in eruditionem universalem*; Wolf entitled his, from 1785 onwards, *Encyclopaedia philologica*. He repeated

---

[1] See above, p. 161.   [2] See above, p. 168; Sandys iii 59.

this lecture eighteen times, we are told, giving it its final form after the defeat of Jena, when the university was forced to close down, with the new heading: 'Darstellung der Altertumswissenschaft nach Begriff, Umfang und Zweck'. He even published it as the first article of his new periodical 'Museum der Altertumswissenschaft'.[1] J. J. Scaliger[2] had been the first to conceive the idea of an all-embracing scholarship of the ancient world, and Winckelmann was the first to understand Scaliger and to follow him in principle.[3] Wolf, who was dependent on Winckelmann in so many ways, took the idea over from him, not from Scaliger,[4] and coined the term for it that has remained in use ever since; as we shall see, it is an apt description of a particular form of the *philologia perennis* in the German nineteenth century.

Wolf was not only an influential writer, but also an effective organizer.[5] His favourite creation was his philological 'Seminar', intended specially for the training of classical teachers. Göttingen and Leipzig also had classes as well as lectures, but since they were occasional and temporary they are not to be compared with Wolf's permanent and methodical creation at Halle. After the catastrophe of Jena in the Napoleonic wars the university closed down, and in 1810 when Humboldt founded the new university of Berlin he tried to use Wolf's abilities and experience for his new foundation. Unfortunately Wolf's strength was exhausted; but the foremost pupils of his seminary in Halle, Böckh and Bekker, became the glories of the new university.[6]

The great German scholar whose activity we have tried to describe was an exceedingly difficult character. For that reason not only was he personally disliked in his own time and afterwards, but the achieve-

---

[1] 1 (1807) 1 ff. reprinted in *Kleine Schriften* II 808 ff.
[2] See above, p. 118.   [3] See above, p. 170.
[4] I have not found a direct reference to Scaliger in Wolf's writings, but I might have missed it.
[5] Cf. above, p. 173.
[6] A number of his books are quoted in the text; his articles are collected in *Kleine Schriften*, in 2 volumes edited by Bernhardy (1869). Selected bibliography in Sandys III 60 f. with reference to Goedeke, *Grundriss* VII² 807–11.—The most important publication is F. A. Wolf, *Ein Leben in Briefen*. Die Sammlung besorgt und erläutert durch Siegfried Reiter I 1779–1807; II 1807–1824; III Erläuterungen (Stuttgart 1935). Cf. my review, *Gnomon* 14 (1938) 401–10. Ergänzungsband I Die Texte (Briefwechsel Wolfs mit Bekker, pp. 1–86, Briefe Wolfs an verschiedene Adressaten pp. 87–161) 1956. The promised commentary had not yet been published.—On his life see 'Entwurf einer Selbstbiographie' in Reiter's edition of the letters II 337–45; 'Entwurf einer zweiten Selbstbiographie', Ergänzungsband I, ed. by Sellheim, pp. 162–6.—W. Körte (Wolf's son in law), *Leben und Studien F. A. Wolfs, des Philologen* (2 vols., 1833). M. Bernays, *Goethe's Briefe an F. A. Wolf* (1868) with introduction.—Mark Pattison, 'F. A. Wolf', *Essays* I (1889) 337–414 (actually a review of J. F. J. Arnoldt, *F. A. Wolf in seinem Verhältnis zum Schulwesen und zur Pädagogik*, 2 vols. 1861/2, but one of the best appreciations of Wolf's person and work).

ments of his scholarship have also been unjustly depreciated. The scholar most conspicuously guilty of this in recent times was Wilamowitz, though even he showed some awareness of Wolf's merits as a textual critic and as a historian. No one knew Wolf better than Wilhelm von Humboldt; he saw his greatness and his deficiencies and, in contrast to Wilamowitz, remained just to him, as even Goethe, Wolf's greatest friend, could not afford to be.

XV

WOLF'S YOUNGER CONTEMPORARIES AND PUPILS

DIFFICULT though Wolf might be, no classical scholar could call so many great contemporaries his personal friends. He also had eminent pupils well qualified to continue his work. It has even been said that his greatest work was his pupils;[1] but that is going too far; his *Prolegomena* and the related studies and texts were beyond doubt his greatest achievements.

Before turning to Wolf's pupils and successors, we must mention two younger contemporaries of his who were not dependent on him as personal pupils. Of all the German classical scholars Gottfried Hermann in Leipzig (1772–1848) was the nearest in approach to Bentley. Herman was not simply a Hellenist like the other scholars in Germany who followed Winckelmann, but also a great Latin scholar[2] in the field of early language and metrics, as Bentley had been. He may be compared with Bentley even in the respect that behind his critical operations there was a clear conception of the ideal classical work, founded on the belief that he, as a scholar, knew what the poet ought to have said ('quid debuerit poeta dicere'). As early as his *De poeseos generibus* of 1794[3] he was employing a quite definite system of aesthetic and critical terminology. It might well be possible to show that the terms used in his grammatical, metrical, and critical writings were derived from a certain section of Kant's transcendental analytics.[4] Hermann's objection to the method of the English scholars was that they were content simply to derive general rules from series of examples; these empirical statements could never be sufficient, Hermann argued; it was always necessary to investigate the law which was manifested in the examples, to ask, for instance, what was the reason for the so-

---

[1] M. Pattison, *Essays* I (1889) 337 ff. =*North British Review*, June 1865.

[2] Cf. E. Fraenkel, 'The Latin Studies of Hermann and Wilamowitz', *JRS* 38 (1948) 28–34.

[3] G. Hermann, *Opuscula* I (1827) 20 ff.

[4] Perhaps a classical scholar better acquainted with the *Kritik der reinen Vernunft* will be able to give chapter and section.

called 'lex Porsoni'. With this logical precision he combined a very fine sense of Greek idiom. In his editions of Greek tragedies he may be said to rival Porson and the Porsonians or even surpass them, especially in Aeschylus. His complete edition was not published until 1852 after his death; but he had been preparing it throughout his long life. Wilhelm von Humbolt was in touch with Hermann for many years, constantly improving and changing his own translation of the *Agamemnon*. Shortly after the battle of Leipzig in 1813 the two men were walking over the battlefield, discussing the text of some passages of this play. Suddenly Humboldt, who had come from a diplomatic mission, interrupted the discussion and said to Hermann: 'As you see here, empires perish, but a good poem lives for ever.'[1] Humboldt's whole personality is revealed in this momentary *aperçu*.[2]

Hermann did not confine himself to the dramatists. Neither epic nor lyric poetry lay outside his scope, and in his *Orphica* (1805) he even gave a full history of the epic hexameter, in which he published some important discoveries. But, since Wilamowitz, whose peculiar aversion against Wolf we have remarked, repeatedly emphasized that this part of the *Orphica* was Hermann's highest accomplishment, it must be put on record that the inspiration for his historical researches clearly came from Wolf; and when Hermann finally began to analyse the text of the Homeric hymns, of Hesiod, and finally of the *Iliad* and *Odyssey*, he was cultivating the same field that had been opened by Wolf's *Prolegomena*. Hermann disliked every kind of post-Kantian idealism, and especially romanticism. In him there lived the belief that the Greek classics had been of perfect beauty, that he could exactly define what was beautiful, and that therefore he could explain or restore the texts of the classics.

The second of Wolf's younger contemporaries was of a quite different type and sometimes an opponent of Hermann: Friedrich Gottlieb Welcker (1784–1868). He became the most intimate friend of Humboldt, when the latter was Prussian ambassador to the Vatican, and he met there the ingenious Danish archaeologist Johann Georg Zoëga (1755–1809), who was a friend of Thorwaldsen and the author of *Ancient Roman Bas-reliefs* and writings on ancient religion. Welcker had an even profounder feeling for Greek religious myth, and he was the first after Winckelmann to combine a true knowledge of poetry with a deep

---

[1] See Leitzmann, *Festschrift für Judeich* (1929) p. 236; cf. *Wilhelm und Caroline von Humboldt in ihren Briefen*, hrsg. v. A. v. Sydow (1906–) IV 149 and 197.

[2] Which is characteristic also of this whole unique period in German history.

understanding of art.[1] His highest purpose was to write a coherent series of books on Greek religion, Greek art, and Greek poetry, and indeed he executed a very large part of his plan in the long series of his writings. After having finished the three volumes of his *Griechische Götterlehre* (1851–63), he dictated as an old blind man of eighty-four his last essay on the 'serenity and beauty of Greek religion'.

His outstanding knowledge and his rare understanding of the Greek genius as a whole enabled him to reconstruct lost parts of Greek poetry: the *Epic cycle*, that is the Trojan epics other than the *Iliad* and *Odyssey* (2 volumes, 1835–49), the lost Aeschylean trilogies (2 volumes 1824–6), and the lost tragedies of all the other dramatists, *Greek Tragedies in relation to the Epic Cycle* (3 volumes 1839–41). It is very easy to blame him for being too imaginative, and Hermann's sober scepticism will always be a sound antidote, as we have learnt again and again from recently published tragic papyri.[2] But it was only because Welcker saw the patterns of Greek legends and recognized their leading ideas that he rediscovered the law of the tragic trilogy in Aeschylus. The articles he published on Greek lyric poetry, especially on the religious lyrics,[3] are still the most adequate publication on these delicate fragments of Greek poetry.

Wolf's and Hermann's pupils and their pupils in the following generation began to form schools. In Welcker's case the formation of a school is inconceivable; he became immortal only because of his achievements. Rivalry between the different schools was inevitable; the winner was that of 'Altertumswissenschaft' founded by F. A. Wolf and propagated by one of his foremost pupils, August Böckh, who nevertheless strongly criticized Wolf's 'Darstellung' and vehemently attacked him personally.[4] 'Altertumswissenschaft' is one particular form of the *philologia perennis*; Hermann, Welcker, and Bekker cannot be counted among its exponents, and there were always classical scholars inside and outside Germany opposed to it. Even Wolf's favourite pupil went his own way, Immanuel Bekker (1785–1871); he had no enthusiasm for the historical Altertumswissenschaft, but devoted his life to the editing of an astonishing quantity of ancient

---

[1] 'Philologie ohne Kunstbegriff nur einäugig', proclaimed Goethe for the generations after Winckelmann.

[2] It is useful to compare Hermann's and Welcker's old reconstructions with the text of the new papyri.

[3] See especially his *Kleine Schriften*, vols. I and II (1844).

[4] F. K. J. Schütz, *Chr. Gottfr. Schütz, Darstellung seines Lebens nebst Auswahl aus seinem literarischen Briefwechsel* I (1834) 13, Böckh to Schütz, 9 Oct. 1812, 'Infamie', 'schuftige Rolle'.

texts. Starting with reviews of Heyne's and Wolf's Homeric publications,[1] he went on first to the texts which Wolf had planned to edit, Homer and Plato, and then to a considerable number of others. He collated more than 400 manuscripts in all the European libraries and published nearly a hundred volumes of ancient and Byzantine Greek texts, which included the standard edition of Aristotle in the four quarto volumes of the Berlin Academy and *Anecdota Graeca* (3 vols. 1814–21); but he also published some Latin and even French texts. In the quantity of his productions he was a sort of Stephanus of the nineteenth century; the standard of quality had of course considerably improved in the meantime. Bekker held an 'ordinary' professorship in the university of Berlin for sixty years, but he was famous for the skill with which he managed to avoid the burden of lecturing during almost the entire time, as Porson had done.[2]

Böckh, who was born in 1785 at Karlsruhe, and died in Berlin in 1867, was the pupil at Halle not only of Wolf in classics, but also of Schleiermacher in philosophy. His own bent of mind was a truly philosophical one, and that distinguished him from the other classical scholars. He did not espouse any particular system, but was indebted generally to the idealistic philosophy of his time and to the romantic movement, with its lively sense of history. His *Enzyklopaedie und Methodologie der philologischen Wissenschaften* is the document of this rare combination. Where Wolf had given only a sketch of Altertumswissenschaft he tried to give a systematic structure and to replace merely practical rules of interpretation with fixed laws based on a comprehensive theory of hermeneutic.[3] It is not surprising that he had no real followers in this dangerous general undertaking. On the other hand, many tried to follow his example in the publication of very specialized books and articles on metre and metrology, inscriptions, finances, and astronomy, or on Pindar, tragedy, or Plato. Behind all these studies of single topics there was always the unifying idea of the knowledge of the ancient world as a whole. At the same time he was convinced like Wolf and Humboldt under Winckelmann's influence that the fundamental ideas of the creative human mind and the first patterns of the beautiful had originated in the achievements of the Greeks, and that we should emulate them. It is obvious that Böckh's universal historical conception was different from Hermann's conscious limitation to the critical

---

[1] *Homerische Blätter* (=*Carmina Homerica* I. Bekker emendabat et annotabat, vols. III–IV, 1863–72).

[2] See above, p. 160.   [3] Cf. above, p. 93.

restoration and interpretation of texts. It is usual to read in histories of classical scholarship that from this moment onwards two rival schools persisted through the nineteenth into the twentieth century, one concerned with words, the other with things,[1] and that the activity of scholarship for several generations consisted mainly in polemics between the two schools.

Possibly this simple scheme is an invention of Conrad Bursian[2] in his *History of Classical Philology in Germany*, published in 1883 as a volume of more than 1,300 pages. As a matter of fact, classical scholarship continued to develop during the nineteenth, as in previous centuries, in step with the general intellectual movements, although owing to its tradition of 2,000 years it was less dependent on the changing spirit of the ages than other provinces of learning.[3]

[1] The quarrel between the Alexandrians and the Pergamenes might be regarded as a parallel in antiquity. Cf. *History* [1] 172 and 237.
[2] I have not come across it in any earlier book.
[3] Cf. R. Pfeiffer, *Philologia Perennis* (1961) p. 22.

XVI

THE BEGINNING OF THE NINETEENTH CENTURY. GERMAN ALTERTUMSWISSENSCHAFT FROM NIEBUHR TO DROYSEN

WINCKELMANN's ideas and writings were decisive for the future of classical scholarship. Its development in the course of the nineteenth century confirms this claim. For in Germany the dominant influence was that of 'Altertumswissenschaft', as F. A. Wolf called the approach to classical studies that embraced all aspects of the ancient world including religion. This had been the conception of J. J. Scaliger, though Winckelmann was the first to recognize its importance.[1] To make it known to a wider circle was one of the achievements of Niebuhr, who was much more than a specialist in Roman history.

Barthold George Niebuhr[2] (1776–1831) was born at Copenhagen, the son of Karsten Niebuhr, the famous traveller in the Near East, and studied classics and history at Kiel (1794) and Edinburgh. In 1806 he became a Prussian civil servant, and then in 1810 as a member of the Berlin Academy he began to lecture on ancient, especially Roman, history in the university of Berlin, just founded by Wilhelm von Humboldt. The immediate outcome of his lectures on Roman history at Berlin was his book entitled *History of Rome*.[3] But in the tradition of

[1] See above, p. 118. I prefer the term Neohellenism to New Humanism. We saw that in the humanism of the fifteenth to the eighteenth century the religious problem played a dominant part. This is no longer the case in the eighteenth and nineteenth centuries. Winckelmann and Humboldt had no interest in religion and especially not in Christianity; this was decisive for the change.

[2] *Lebensnachrichten über B. G. Niebuhr aus Briefen desselben* (3 vols., 1838/9). J. Classen, *B. G. Niebuhr. Eine Gedächtnisschrift zu seinem hundertjährigen Geburtstag* (1876); H. Nissen was expected to write *the* biography, but published only a short life in *ADB* 23 (1886) 646 ff. The letters were edited by D. Gerhard and W. Norvin (2 vols., 1926–9). On German writers about history see F. Schnabel, *Deutsche Geschichte im 19. Jahrhundert* (4 vols., 1929–37) written in a true humanistic spirit, and G. P. Gooch, *History and Historians in the Nineteenth Century* (2nd ed. revised 1952), on Niebuhr pp. 14 ff.

[3] *Römische Geschichte* I (1811) II (1812). A. Momigliano, 'G. C. Lewis, Niebuhr e la critica delle fonti', *Rivista storica italiana*, 64 (1952) 208–21 on Niebuhr's method and his relation to England.

Scaliger and Wolf he set his Roman history against the background of the history of the ancient world as a whole. Niebuhr was a figure of some complexity, with experience of the practical service of the state as well as the literary tradition of the state university. The main source for early Roman history was, of course, Livy; and just as textual criticism had started with Lorenzo Valla's work on Livy's text,[1] so now it was from Livy that historical criticism was born. A most important factor in Niebuhr's approach was his passionate feeling for the civilization of his native country, its origin, and customs. He believed that he could understand the civilization of early Rome from the analogy[2] of the peasant communities of his northern surroundings, the Dithmarschen, which he knew and loved so well.

Although he felt quite happy in 'Moor und Heide' amongst free peasants, he very much enjoyed his circle of close friends in Berlin; on one of them, Friedrich Karl von Savigny (1769–1861), he relied for the study of Roman law,[3] which Savigny had just established on new foundations. He also liked to travel through German and Italian cities, in the libraries of which he made surprising discoveries of Latin texts, the most spectacular of these being the palimpsest of Gaius' *Institutiones* in the Capitular Library of Verona; in a Vatican manuscript he found fragments of some speeches of Cicero; he made contributions to Mai's edition of Cicero's *De re publica* from a Vatican palimpsest; on his return through Switzerland to Germany in 1823 he identified the contents of a palimpsest in St. Gallen and produced the first edition of Merobaudes. But his years as Prussian ambassador in Rome (1816–23) were disappointing, as he utterly disliked the Rome and Italy of his day.

Wilhelm von Humboldt (1767–1835) had personal and literary relations with all the leading classical scholars of his time,[4] and he felt quite at home while serving in Rome as Prussian ambassador to the Vatican (1802–8). A deeper contrast than that between Niebuhr[5] and Humboldt in Rome can hardly be imagined. Niebuhr's attitude indicates the end of the German period of Neohellenism. Humboldt,

---

[1] Cf. above, pp. 36 ff.

[2] Analogies play an important part in Niebuhr's argumentation.

[3] On earlier legal studies of humanists and classical scholars see above, pp. 86 (Italy) and 101 (France). On Savigny see Adolf Stoll, *F. K. von Savigny* (3 vols., 1927–39), esp. 138 ff.: 'Der junge Savigny'.

[4] See above, p. 176 (F. A. Wolf und G. Hermann), 179 (F. G. Welcker); Niebuhr's depreciatory remarks on Welcker, his colleague in Bonn, whom he was unable to understand, are characteristic.

[5] It seems that he became a member of the Prussian academy through the influence of Humboldt (Wilamowitz, *Geschichte der Philologie*, 1921, p. 53).

on the other hand, said in his hymnic praise of Rome: 'Wie Homer sich nicht mit den andern Dichtern, so läßt sich Rom mit keiner andern Stadt vergleichen', and Goethe quoted this sentence in his memorial book on Winckelmann.[1] It was on the recommendation of Humboldt that Franz Bopp (1791–1867) was given a professorship of 'Orientalische Literatur und allgemeine Sprachkunde' in 1823 at the university of Berlin; he had in 1816 (at the age of twenty-five) laid the foundation of 'Indogermanische Sprachwissenschaft'. Humboldt's earlier writings had been educational and historical, but after retirement from his high diplomatic and ministerial offices in 1819 he concentrated on the study of language. His highest and most original achievements were in this field.[2] As his interests were world-wide, he was able to recognize the significance of Bopp's discovery of the relationship of Sanskrit to the Greek, Latin, Persian, and Germanic languages.[3]

It may seem paradoxical to group Jacob Grimm (1785–1863) with Humboldt. For Grimm regarded individual observation as 'die Seele der Sprachforschung'[4] and, unlike Humboldt, hardly ever entertained theories. Yet, even if not explicit, they were present in his mind and influenced his grammatical writings, and for that reason we may justifiably regard him as one of Humboldt's school.[5]

Philipp Buttman (1768–1829), born one year after Humboldt, came of a family of French emigrants (Boudemont). He remained a pure grammarian without any inclination towards the philosophy of language or comparative linguistics. After studying at Göttingen and Strasbourg he became a member of the Academy in Berlin and head of the library there.[6] As early as 1792 he published his first small Greek grammar, which in the course of time was enlarged to the 'Complete Grammar' of 1819–27. His *Lexilogus*,[7] one of the most influential books on Homeric

---

[1] Cf. above, p. 169.

[2] On H. Steinthal, *Geschichte der Sprachwissenschaft bei den Griechen und Römern* dedicated to Böckh in 1863 (2nd ed. 1890–1) see *History* [1] pp. xviii and 59.3.

[3] Bopp's proof was based on the observation of the common inflexion of the verb in these languages, and was in the end generally accepted; cf. S. Lefmann, *Franz Bopp, sein Leben und seine Wissenschaft* (2 vols. and 'Nachtrag', 1891–7).

[4] *Deutsche Grammatik* I² (1822) p. vi.

[5] Though inclined to this view, I was not convinced until I read Brigit Beneš, *W. von Humboldt, Jacob Grimm, August Schleicher. Ein Vergleich ihrer Sprachauffassungen* (Diss. Basel 1958) pp. 41 ff. (I owe the knowledge of this dissertation to my colleague Meinrad Scheller). In general see J. Dünninger (above, p. 64 n. 1) on the origin of German scholarship.

[6] See Konrad Kettig in *Bibliothek und Wissenschaft*, hrsg. von S. Joost, v (1968) 103 ff.

[7] *Beiträge zur griechischen Worterklärung hauptsächlich für Homer und Hesiod* (1st ed. 1818, followed by many later editions).

language, belongs to the great tradition of Homeric research in Germany stemming from F. A. Wolf.[1]

After Homer Plato had been Wolf's favourite author. His friend Friedrich Schleiermacher (1768–1834)—they met at Halle in 1804—produced a translation and helped to promote Platonic scholarship. But the credit for proposing (in 1798) the idea of a complete translation into German is due to Friedrich Schlegel[2] whose interests were more philosophical[3] than those of the other romantics.[4] In his early years as a student of classics and an admirer of F. A. Wolf[5] Schlegel had devised his literary categories, which had an enormous success far beyond classical scholarship. By means of them he demonstrated the superiority of the creations of the Greeks in literature and art and at the same time threw light on their historical position and influence on the future.

Karl Otfried[6] Müller (1797–1840) will always live on as the radiant figure of a happy young scholar whose life came to a premature end in Greece, in the country with which he was desperately in love. He still belonged in spirit to the age of Winckelmann; born in the Silesian town of Brieg, he studied first at Breslau, and then in Berlin came into close contact with the illustrious circle of classical scholars there, led by F. A. Wolf, whom he found personally repellent, but whose writings had a permanent influence on him, especially in his reverence for history.[7] He was more beloved by Böckh than any other of his numerous pupils (1816/17). Müller's complete local history of Aegina,[8] *Aegineticorum liber* (1817), proved him to be equally familiar with the monumental and with the literary sources, and in 1819 he was appointed professor of classical Altertumswissenschaft in Göttingen. The most original of his various publications he entitled *Prolegomena zu einer wissenschaftlichen Mythologie* (1825), obviously in imitation of the title of Wolf's most famous book on Homer. The main thesis of this work on

---

[1] Cf. above, pp. 175 f.

[2] F. Schlegel, *Kritische Ausgabe seiner Werke*, von E. Behler, Hans Eichner u.a. (München–Zürich 1958–) and many separate publications.

[3] F. Schlegel, 'Philosophie der Philologie', *Logos* 17 (1928) 1 ff.

[4] Schleiermacher in a letter to Böckh, dated 18 June 1808 (*Mitteilungen aus dem Litterarurarchive in Berlin* N.F. 11 (1916) 26) mentioned F. Schlegel's casual remark in conversation with his Berlin friends who in his opinion should undertake the complete translation as a joint enterprise.

[5] On F. Schlegel's most enthusiastic review of Wolf's *Prolegomena* see above, p. 175.

[6] He added this second Christian name to Karl in 1819, *Briefe* hrsg. u. erläutert von S. Reiter 1 (1950) 10 ff. (at the advice of Buttmann, said Richard Foerster, in his paper *Otfried Müller*, Breslau 1897).

[7] See above, p. 174 'reverenda est historia'.

[8] Bibliography with exact titles see *Briefe* 11 (1950) pp. ix ff.

mythology[1] was that Greek myths contain the earliest history of the Greek tribes. This, both in itself and because of the valuable controversy it provoked, was Müller's most important contribution to scholarship.

The most vigorous opposition to this 'historical' theory of mythology was expressed by Friedrich Creuzer (1771–1858), who, under the influence of Joseph Görres's mysticism, presented the religious ideas of the ancient world in the four volumes of his *Symbolik* (1810–12, later editions 1819–23 and 1837–43 = *Deutsche Schriften* Abth. 1 Bd. 1–4);[2] his complete edition of Plotinus, published by the Clarendon Press in 1835, was connected with these studies.

Müller's belief in the historical importance of myth is a sort of parallel to Niebuhr's attitude towards early Roman civilization; in both cases the influence of Romanticism is evident. There is another parallel in so far as K. O. Müller also went to England, where he stayed for many of the remaining years of his life. There in 1836 he began to write his most successful book, the *History of the Literature of Ancient Greece*[3] commissioned by the London Society for the Diffusion of Useful Knowledge; it was still unfinished when he died in August 1840 in Athens and was buried on the hill of Colonos.

His *Aeginetica* was followed by three prehistoric works. *Orchomenos und die Minyer* (1820) dealt with the prehistory of Boeotia; *Die Dorier* (2 volumes, 1824) was more an impressive hymn on the excellence of everything Doric than a narration of history; and in 1828 his *Etrusker* won the prize in Berlin, surprisingly for a book in which he was entering a new field.

In the last years of K. O. Müller's life Heinrich Ludolf Ahrens (1809–81) started to publish his work on the Greek dialects.[4] This

[1] On earlier mythography since the Renaissance see above, pp. 20 ff. with notes thereto.

[2] It may be time for a monograph on J. Görres and classical scholarship in general, as it is difficult for a scholar examining occurrences of individual detail to do justice to a totally uncritical polymath like Görres.

[3] K. O. Müller, *History of the Literature of Ancient Greece*, 2 vols., translated from the German manuscript by G. C. Lewis (London 1840–2). *Geschichte der griechischen Literatur bis auf das Zeitalter Alexanders*, hrsg. von Eduard Müller, 2 vols. (Breslau 1841, 2nd ed. 1857, 3rd ed. with notes and additions by E. Heitz, Stuttgart 1875–6). *A History of the Literature of Ancient Greece*, translated from German MS. of K. O. Müller by Sir G. C. Lewis and J. W. Donaldson, continued from German MS. by J. W. Donaldson, 3 vols. (London 1858). K. O. Müller, *Histoire de la littérature grecque jusqu' à Alexandre le Grand*. Trad. et annotée et précédée d'une étude sur O. Müller et sur l'école historique allemande par Karl Hillebrand (2 vols. 1865, 2nd ed. 3 vols. 1866, 3rd ed. 3 vols. 1883). Müller's papers and reviews, of which the variety of subjects is astonishing, were collected and published by the author's brother Eduard Müller (2 vols., 1847/8). G. Bernhardy, *Grundriss der griechischen Litteratur* (3 vols., 1836), unfinished, is only a dry register of titles and dates.

[4] *De Graecae linguae dialectis* (2 vols. 1839). See E. Fraenkel Aesch. *Ag.* 1 (1950) 54 ff.

would not have been possible without the foundation laid in Müller's historical writings (he had been Ahrens's teacher at Göttingen after 1826), and indeed it must be regarded as one of their most important consequences. The study of Greek dialects has not been neglected in ancient or modern times; in the French Renaissance universalists such as Henri Étienne and Claudius Salmasius devoted attention to them. But it was Ahrens who by his methodical treatment advanced dialect-ology to a special branch of classical scholarship which is still today treated on the lines he established (not always to the satisfaction of progressive linguists). He dedicated the second volume, *de dialecto Dorica*, to Lachmann. His critical edition of *Theocritus* (with the scholia) and the minor *Bucolic poets* (1855–9) was acknowledged as a masterpiece by the greatest later editor, Wilamowitz (1913).[1] Ahrens held a number of scholastic appointments before he settled down as a Director of the Lyceum of Hannover from 1849 to 1879, where Raphael Kühner (1802–78), the foremost teacher of the traditional Greek grammar, had published his manual in 1844–5. Ahrens's *Kleine Schriften* (1891) contain famous articles on classical authors and on linguistics; but he is here intentionally placed with the successors of K. O. Müller, not with the linguists who followed Humboldt. Even more than Bopp, it was J. Grimm who influenced his ideas.

In the preceding chapters the historical element was very strong, but there was no true historian in Germany after Niebuhr. J. G. Droysen (1808–84)[2] certainly was a historian, but even he started as a pupil and friend of F. G. Welcker with the study of Aeschylus and Aristophanes; for in contrast to Niebuhr he had a genuine appreciation of great poetry and of Welcker's scholarship. However, from 1840 onwards Droysen occupied chairs of history, including modern history, first in Kiel, then from 1849 in Jena, and from 1859 in Berlin. In Berlin he enjoyed very friendly literary and personal relations with the Mendelssohn family, especially the musician Felix Mendelssohn-Bartholdy,[3] who occasionally set lyrical poems by him to music;

---

[1] In Schol. [Theocr.] IV 16a everyone has accepted his emendation πρώκιον (*vocabulum novum*, as πρῶκαι is in the text of Theocritus) except Paul Maas, *Kleine Schriften* (1972) pp. 210 f., who constantly defended the reading of the manuscripts προίκιον, which does not fit the meaning of the scholion. In [Theocr.] VIII 91 Ahrens's ἀμαθεῖς is equally convincing (γαμηθεῖσ', γαμεθεῖσ' codd.), and there are many other examples.

[2] G[ustaf] Droysen [Sohn], *Johann Gustav Droysen* I: Bis zum Beginn der Frankfurter Tätigkeit (1910). J. G. Droysen, *Briefwechsel*, ed. by R. Hübner, Deutsche Geschichtsquellen des 19. Jahrhunderts 25–6 (1929). On Droysen's concept of history see B. Bravo, *Philologie, histoire, philosophie de l'histoire. Étude sur J. G. Droysen, historien de l'antiquité* (1968).

[3] Felix Mendelssohn–J. G. Droysen, *Briefe*, hrsg. v. C. Wehmer (1959).

conversely Droysen used Mendelssohn's discoveries, especially with regard to J. S. Bach, for a better understanding of Aeschylus' tragedies. Droysen's Aeschylean trilogy exercised a certain influence on the conception of Richard Wagner's Nibelungen tetralogy.

Droysen's later historical works were much admired for their perfect literary prose. Only a poet who had translated Greek tragedies and comedies into German verse could produce such works of prose as his *Alexander* and his *Hellenismus*.

These volumes on Alexander and his successors (1836) laid their stress not on the past of the 'classical' Greek world, but on the post-classical centuries. The different character of these centuries, their new significance, and their own greatness were demonstrated and explained. Droysen recognized in them a period with its own 'historical principle' (as he said with Hegel), an epoch in which the Greek genius progressed towards new achievements.[1] He was particularly lucky in finding a new term for this epoch, 'Hellenismus'. He may have read the expression 'Hellenistische Sprache' in Buttmann's Greek grammar (1819)[2] where it was described as a modern usage ('neuerer Sprachgebrauch'). Droysen disagreed claiming in the Vorrede of his *Geschichte des Hellenismus* (I, 1836, p. vi): 'Es ist aus dem Altertum überliefert, die Sprache jener westöstlichen Völkermischung mit dem Namen der hellenistischen zu bezeichnen.' There is, however, no such ancient tradition. One can try to reconstruct it from ancient passages where ἑλληνίζειν, ἑλληνιστής, ἑλληνισμός occur.[3] But this hypothetical reconstruction is certainly not sufficient to justify talk of an 'ancient tradition'. The disagreement between Buttman and Droysen is insoluble as long as no evidence turns up for the ancient usage. But there is sufficient evidence for the existence of the modern usage before the nineteenth century, which neither Buttman nor Droysen seems to have known. In the circle of J. J. Scaliger's pupils in the early seventeenth century there were passionate discussions about the existence and the meaning of 'lingua Hellenistica'[4] as a special Greek 'dialect' used in the biblical writings.

[1] Quoted from 'The Future of Studies in the Field of Hellenistic Poetry', *JHS* 75 (1955), reprinted in *Ausgewählte Schriften* (1960) p. 151.

[2] *Ausführliche griechische Sprachlehre* I (1819) 7, n. 12 'Aber auch die ungriechischen Bewohner . . . fingen nun an griechisch zu sprechen (ἑλληνίζειν) und ein solcher griechisch redender Asiat, Syrer usw. hieß daher ἑλληνιστής. Hieraus ist der neuere Sprachgebrauch entstanden, daß man die mit vielen ungriechischen Formen und orientalischen Wendungen gemischte Schreibart von Schriftstellern dieser Art die Hellenistische Sprache nennet.'

[3] See A. Debrunner, 'Geschichte der griechischen Sprache II: Grundfragen und Grundzüge des nachklassischen Griechisch', *Sammlung Göschen* 114 (1954, 2nd ed. 1969) 10 f.; but this reconstruction is not at all convincing.

[4] See especially Claudius Salmasius, *De lingua Hellenistica commentarius* (1643), and an

In the course of the nineteenth century, especially during the second half, the study of ancient languages began to flourish once again beside that of ancient history. It was characteristic of this period that Latin was able to regain the place beside Greek that it had occupied before Winckelmann. Karl Lachmann (1793–1851) seems to have led the way here; his 'pater studiorum', as he called Gottfried Hermann, was a scholar and teacher of equal distinction in Latin and in Greek. Of Lachmann's Latin texts, Propertius (1816, 2nd edition 1819 with Catullus and Tibullus), and Lucretius (1850), the Lucretius is regarded as exemplary. From the Latin elegiac poets he went on not only to Lucretius, but to the great epic narrative poets outside the classical world. Mommsen called him the greatest master of language.

But Lachmann's name is famous above all for his method of textual criticism,[1] in which 'recensio' of the manuscripts led to the so-called 'archetypus', and if examination of the manuscripts did not result in a convincing original reading, 'emendatio' was necessary. He developed this method in his work on Latin poets, and brought it to such a level of accomplishment in the study of the manuscripts of the New Testament that he was able finally to discredit the 'textus receptus'.

This part of our History, which started with Petrarch, has followed through all their changes the leading ideas first of Italian humanism and then of German Neohellenism. At the point we have reached, in the middle of the nineteenth century, there was a definite break. No longer was humanism to be the driving force. A new world full of contrasts emerged, of which Theodor Mommsen (1817–1903) was the greatest scholarly figure. While still himself an admirer of Humboldt's idea of the state, he yet did more than anyone else to further the forces of historicism and realism.

anonymous writing (wrongly attributed to Salmasius in the bibliographies) *Funus linguae Hellenisticae* (1643), in which the existence of a 'Hellenistic language' at any time is denied.

[1] See S. Timpanaro, 'La genesi del metodo del Lachmann', *Biblioteca del Saggiatore* 18 (1963), German translation with corrections and additions (1971). Paul Maas, *Textkritik* (4th ed. 1960) does not even mention the name of Lachmann.

GENERAL INDEX

*indicates a specially extensive reference

academies of the Italian Renaissance: 56 f.
— Roman academy, founded by Pomponio Leto: 51 f.
Acciaiuoli, Niccolò: 23.7
Acton, Lord, and Döllinger: 77
Aegina, complete local history, given in K. O. Müller's *Aegineticorum liber*: 186
Aeschylus, Codex Laur. Mediceus XXXII 9: 48
— recension of Triclinius: 111
— Dorat's textual criticism: 104, 160
— Turnebus's and Robortello's editions: 111
— Canter's edition: 125
— Th. Stanley as editor and critic of A.: 144, 161
— G. Hermann's edition: 179
— Robortello's first edition of the Scholia: 136
— *Prometheus*, read by Dorat to his pupils: 105
— *Agamemnon*, first complete edition by P. Victorius: 111, 136
— Porson on *Ag.* 1391 f.: 161.2
— — translated by W. v. Humboldt: 179
— — Porson's textual notes: 159 f.
— Welcker's reconstruction of the lost trilogies: 180
— Droysen's study of A.: 188 f.
Aesop, translations of smaller excerpts into Latin by Valla: 38.5
— translated and introduced by Luther: *91
— admired by Sir William Temple: 150
Agricola, Rudolf: *69 f.
— and Italy: 64
— and Erasmus: 71.2
— revived the genre of 'loci': 92
Agrigentum: 170
Agustín, Don Antonio: 95
Ahrens, Heinrich Ludolf: 187 f.
Albrecht V, Duke of Bavaria: 140
Alcalà, university, Greek studies: 65 f.
— — and Sepulveda: 94
Alciato, Andrea: *86, 95, 100
Aldine Academy: 56
— and Erasmus: 56, 73
Aldus Manutius, *see* Manutius, Aldus
Alexander the Great, Droysen's work on him and his successors: 189
Alexander of Villedieu: 52
Alexandria, centre of scholarship: 50
— library: 50, 174

— modern parallel of the quarrel between the Alexandrians and the Pergamenes: 182
— grammarians: 157
— — text of Homer: 157, 174
Alfonso, king of Aragon and Sicily: 35 f., 56, 61
allegorical interpretation:
— Boccaccio's *Genealogie* preserved explanations of myths in Stoic allegorical traditon: 21
— in Salutati's *De laboribus Herculis*: 27
— not practised by Colet: 72
— of Homer by Dorat and his contemporaries: 104.3
'Altertumswissenschaft', conception of: 118, 175 f., 180 f., 183, 186
Ambrogini, Angelo, *see* Politian
Ambrose, Saint, and Petrarch: 12
— edition of Erasmus: 78
Amerbach, family: 83, 86.9
Ammianus Marcellinus, XVIII 2.15, B. Rhenanus's conjecture: 84.1
— lost manuscript from Hersfeld: 85
— ed. by S. Gelenius: 85
Amyot, Jacques, in audience of the *lecteurs royaux* for Greek: 102
— translation of Plutarch: 113, 141
Anacreontea, H. Étienne's first edition: 104, 106 f., 109
analogies in Niebuhr's argumentation: 184
ancient authors, Boccaccio's trust in their infallibility: 22
Anglus, Ioannes Clemens: 109.3
Anthologia Graeca, treated by Politan: 44
— Codex Palatinus 23: 48, 138
— — and Salmasius: 122
— H. Étienne's edition of the *Anth. Planudea*: 109
— Latin translation of the *Anth. Planudea* by H. Grotius: 127
Antioch (Syria): 149
'antiquitates': 133.3
antiquities, interest in: 50 f. (F. Biondo), 95 (A. Agustín); *see also* archaeology
Antoninus, Marcus Aurelius, *ed. princeps* by C. Gesner, not by Xylander: 140.7
— Gataker's commentary: 144.2
Antwerp: 138
Apollonius Rhodius, Codex Laur. XXXII 9: 48, 136
— among Dorat's favourites: 104

Appendix Virgiliana, Boccaccio the first to get hold of parts of: 24
— edition of J. J. Scaliger: 117
Appius Claudius Caecus as translator of Greek maxims into Latin: 28
Apuleius, *Metam.*, commentary of Beroaldo: 55.5
Arabic studies of Reiske: 172
Aratus, among Dorat's favourites: 104
— edition of H. Grotius: 126
'archaeographia': 132
archaeology:
— Buondelmonti's *Descriptio Candiae* the earliest book of archaeological travels: 29
— Peiresc, Spon, and *archaeologia*: 133
— excavations of Herculaneum and Pompeii: 170
— Winckelmann's vision of the discovery of Olympia: 171
— G. Zoëga's work: 179
'archetypus': 190
Aretino, Francesco, completed Valla's translation of the *Iliad:* 38
argumenta: 4
Argyropoulos, Johannes: 87
Ariosto, Ludovico, *Cinque canti:* 21
Aristarchus, written commentaries: 54
— ὑπομνήματα and συγγράμματα: 79
Aristippus, Henricus, Latin version of Plato's *Phaedo:* 14
Aristophanes, *Plutus* 1–269, Bruni's rendering in Latin prose: 28
— — 400–626, Rinucci's paraphrase: 29
— *Clouds*, edition of Melanchthon: 92
— Droysen's study of A.: 188
Aristotle, standard edition of the Berlin Academy: 181
— Petrarch's acquaintance with the *Ethics:* 8
— Bruni's translations of the *Politics* and of the *Ethics* into Latin: 29
— *Poetics*, edited, commented on, and translated by Robortello: 136
— — D. Heinsius's book on the: 129
— Vettori's commentaries: 136
— references to rhythm in prose: 30
Arminians: 127, 144
Arminius, Jacobus: 124
art, understanding of, to be combined with scholarship (Welcker, Goethe): 179 f.
— ancient, history of, Winckelmann's treatment: 168 ff.
— — Scaliger's suggestion that there were four ages of Greek poetry taken up by Winckelmann distinguishing four different styles of Greek art: 119, 170
— — publications of the 'Society of Dilet-tanti' with drawings of works of Greek art: 161 f.
artes liberales: 3
Ascham, Roger: 143
Asconius, commentary on Cicero, discovered by Poggio: 32
astronomy, its renaissance furthered by Bessarion and Regiomontanus: 37
— as basis for historical chronology (Scaliger): 117
— ancient, studied by Böckh: 181
atheism, Bentley's confutation of: 146 f.
Athenaeus, commentary of Casaubon: 121
Athens, acropolis, sculptures, *see* Elgin Marbles
Athos, Mount, Greek manuscripts, brought to Florence by J. Lascaris: 48
Attic dialect and atticistic imitations treated by Bentley: 152
d'Aubignac, François Hédelin, Abbé: 134
Augsburg: 62, 88, 94, 139 ff.
Augustine, Saint, and Petrarch: 10
— edition of Erasmus: 78
— included among the 'classici' by Fonseca: 84.4
— *De civ. dei*, ed. by Vivès: 96
— — XI 18 and Bentley: 146.5
Aurispa, Giovanni: 48 f.
Ausonius, Boccaccio the first to get hold of: 24
— and Politian: 43
Averroists: 14

Bach, J. S.: 189
Bacon, Francis: 114
Baïf, Lazare de: 103
Barbaro, Ermolao: 66
Barbosa, Ayres: 65
Barlaam, teaching Petrarch a little Greek: 14
— library: 14
Barnes, Josuah: 144
— edition of Euripides: 161
Baron, Hans: 27 n. 7, 28 n. 4
Barzizza, Gasparino da, founder of Ciceronianism: 43
— sketchy expositions of a few of Cicero's writings and letters: 54
Basedow, Johann Bernhard: 17
Basle, council of: 59
— the university as home of humanism and classical studies: 60, 83
— printing: 83
— Erasmus in B.: 73, 83
— Cono in B.: 86.9
Beaufort, Henri, cardinal: 33, 61
Beccadelli, Antonio: 57
Bekker, Immanuel: *180 f.

— pupil of Wolf's seminary in Halle: 176
— correspondence with Wolf: 176.6
'Bembismo': 135
Bembo, Pietro: distinguished representative of Ciceronianism: 53
— his virtuosity in Latin verse: 135
Benedictines, French: 156; *see also* Maurists
Beneventan script: 139
Bentley, Richard: *143–58
— called 'tremendous' by Gibbon: 162
— of all the German classical scholars G. Hermann was the nearest in approach to him: 178
Berkeley, George: 144
Berlin:
— Winckelmann in B.: 167
— university: 170, 176, 181, 183, 185, 188
— Academy: 181, 183, 185
— State Library: 185
— circle of classical scholars: 184, 186, 188
Bernays, Jacob, *Joseph Justus Scaliger*: 119.7
— unfinished essay on Gibbon 162.3
— and F. Ritschl: 119.7
Berne, Burgerbibliothek: 132
Bernhardy, Gottfried: 187.3
Beroaldo, Filippo, the elder, commentaries on Latin authors: 55
Bersuire, Pierre: 60
Bessarion, cardinal, treatment of St. John 21: 21 ff.: 37
— and Valla: 37
— presented his manuscripts to the Republic of Venice: 49
— his protegé N. Perotti: 54
— furthering the renaissance of science: 37, 139.4
Beyle, Henri, *see* Stendhal
Bible, Cod. Vat. gr. 1209 (Vaticanus B): 95
— the Polyglot Bible of Alcalà (Complutensis): 65, 94 f.
— the definitive Catholic edition of the Vulgate in 1592: 108
— division of the text into chapters and verses by Robert Étienne: 108
— 'the Scriptures alone the norm' (Luther): 92
— *see al*.o criticism, biblical
— OT, Latin Old Testament in R. Étienne's edition of 1556: 108
— — *Psalms*, Melanchthon's lecture on the Hebrew text: 91
— NT, 'scriptura sacra sui ipsius interpres': 76
— — Greek, the 'textus receptus': 108 f., 156, discredited by Lachmann: 156, 190; cf. also Erasmus's edition and the 'textus receptus': 77

— — the edition of Erasmus: 74, *76 f., 83, 94
— — the editions of Rob. Étienne: 77, 108
— — the Elzevier edition of 1633: 108 f.
— — the edition of J. Mill: 156
— — the edition prepared by Bentley: 156, 173
— — Lachmann's recension of 1831: 109, 190
— — Epistle to the Romans, commentary of Ficino: 57, interpreted by Luther: 90
— — Epistle to Titus, Melanchthon's lecture: 91
— — I John 5:7 and 8 (Comma Ioanneum), genuineness rejected by Erasmus: 148 and Bentley: 160, defended by Travis: 160
Biondo, Flavio, interest in ancient antiquities and monuments: 50
— history of post-ancient times: 51
Blenheim: 147
Blomfield, C. J.: 161
Boccaccio, Giovanni: *20–25
— and the authenticity of the correspondence of St. Paul and Seneca: 40.3
Böckh, August: *181 f.
— and Schleiermacher's hermeneutic: 93
— F. A. Wolf's pupil: 176
— criticized and attacked Wolf: 180
— and K. O. Müller: 186
Böhmer, Johann Friedrich: 131
Boeotia, prehistory of, given in K. O. Müller's *Orchomenos und die Minyer*: 187
Boëthius, plan for translating the whole of Aristotle and of Plato: 28
Bohemia:
— Bohemian prehumanism: 59
— Enea Silvio Piccolomini's *Historia Bohemica*: 60
Boiardo, Matteo Maria, *Orlando innamorato*: 21
Boileau, Nicolas, and the 'Querelle des anciens et des modernes': 134.2
— and Longinus: 137
Bologna: 5.2, 25, 55, 94 f.
Bolt, Robert, *A Man for All Seasons*: 79.6
Bongars, Jacques: 132
Bonn: 184.4
book, importance for scholarship: 47
— collecting, *see* libraries
Bopp, Franz: *185, 188
Boyle, Charles, edition of the letters of Phalaris: 150 f.
Boyle, Robert: 146, 150
Brahe, Tycho de: 114
Bremond, Henri: 130
Bretten: 91
Brie, Germain de: 103

Brühl, count Heinrich von: 171

Bruni Aretino, Leonardo: *27–30

— *Dialogi ad Petrum Paulum Histrum*: 16, 30

— Latin version of Plutarch: 29, 141.1

Brutus, Greek letters, and Erasmus: 75 f.

Buchanan, George: 144, 149

Bucolic poets, ed. by Ahrens and Wilamowitz: 188

Budé, Guillaume de: *101 f.

— charged with diplomatic missions to the popes Julius II and Leo X: 65

Bünau, count Heinrich von: 168

Bunsen, Christian Karl Josias von: 119.7

Buondelmonti, Cristoforo de', *Descriptio Candiae*: 29

— visited the countries of the Byzantine empire: 51

Burckhardt, Jacob, *Die Kultur der Renaissance in Italien*: 18

Burdach, Konrad, Renaissance studies: 18

Burman, Pieter (uncle and nephew): 162; cf. also 182

Bursian, Conrad: 182

Buttmann, Philipp: *185 f.

— on Hellenistic language: 123, 189

— at his advice K. O. Müller added his second Christian name: 186.6

Bywater, I.: 161

Byzantinology, founded by Hieronymus Wolf: 140; *see also* I. Bekker, Hoeschel, Pontanus, Xylander

cabbalistic tradition, studied by Reuchlin: 87

Calepinus, Ambrosius, Latin dictionary: 107

Callimachus, his colloquy with the Muses and Petrarch: 6

— Politian's work on C.: 45 f.

— edition of Tanaquil Lefèvre: 135

— edition of Graevius with the help of Bentley: 152, 162

— *Hymns*, edition of Robortello: 46

— — among Dorat's favourites: 104

— — fragments, Th. Stanley's notes: 144, 152

— — Bentley's collection: 152 f.

— — Bentley's treatment of fr. 21.3: 153.4

— *Lock of Berenice*, translated by Catullus and Politian: 45

— — and Scaliger: 115

— *Hecale*, put together with Catullus' translation of the *Lock of Berenice* by Politian: 45

Calvin, Jean: *100

— influence on Budé: 102

Calvinism: 116 (Scaliger), 120 (Casaubon), 124 (Holland)

Cambridge, classical education: 143

— centre of classical scholars also after Bentley's death: 159

— and Erasmus: 73

— Platonists: 144

— University Library: 120.1

— St. John's College: 145

— Trinity College and Bentley: 147 f., 156.6, 160

Camerarius, Joachim: *139

— pupil of Melanchthon: 94

— and Turnebus: 112

Campana, A.: 32

'canon' as term for selective lists of authors coined by D. Ruhnken: 163.2

Canter, Willem: 104.3, 106, 125

Carducci, Giosuè: 21

Casaubon, Isaac: *120–2

— and the ms. of Julius Africanus: 118

— Henry Savile his host: 144

Cato, Valerius: 105

Catullus, and the Pléiade: 105

— edition of J. J. Scaliger: 117

— edition of Lachmann: 190

— translation of Callimachus' *Lock of Berenice* (c. 66) and Politian: 45

— — and J. J. Scaliger: 115

Cavaillon, Philip of, bishop: 15

Cedrenus, ed. by Xylander: 141

Celtis, Conrad: *63 f.

— edition of Tacitus' *Germania* and the plan of a *Germania illustrata*: 64, 84

— discovered the *Tabula Peutingeriana*: 64, 141

Cesarini, cardinal: 48

Chalcidius, commentary on Plato's *Timaeus*: 14

Chalcondyles, Demetrius: 66

Charles IV, Emperor, and Petrarch: 58

Charles V, Emperor: 79, 95

Charles II, King of England: 122.4

Charles VIII, King of France: 102

Charles IX, King of France: 110, 112

Choniates, Nicetas, first edition by H. Wolf: 140

Christianity, defended against deism by Bentley: 147

— the problem of the relation between antiquity and Ch. discussed in the Italian Renaissance: 58; *see also* 144; *see also* humanism, Christian

Christina, Queen of Sweden: 123, 128

chronology, ancient, study of: 117 f. (Scaliger, Petau), 119 (J. Selden), 137 (C. Sigonio)

Chrysoloras, Manuel, teaching the Greek language in Florence: 27

— correspondence with Salutati: 29

— ᾿Ερωτήματα τῆς ῾Ελληνικῆς γλώσσης: 53

— visit to England: 62

Church, Catholic, and Erasmus: 81
Church Fathers:
— Petrarch made no distinction between the classics and the Ch. F.: 11
— Poggio's study of: 33
— Latin style, and Valla: 37
— editions of Erasmus: 78
— editions of the Maurists: 130 f.
Cicero:
— *humanitas*: 15
— a new oratory modelled on C. in the school of J. Sturm: 84
— and Petrarch: 5 f., 9 ff.
— the 'vetus Cluniacensis' and Poggio: 32
— Bentley's contributions to the text of C.: 155
— *Epistulae*, rediscovered by Petrarch: 9 ff.
— — influence on humanistic epistolography: 26
— — Salutati not the discoverer of the *Ep. fam.*: 26
— Barzizza's sketchy expositions of a few of C.'s letters and writings: 54
— discovery of *De oratore*, *Orator*, and *Brutus* by Gerardus Landriani: 33
— Poggio's discovery of eight speeches at Langres and Cologne: 32
— fragments of some speeches found in a Vatican ms. by Niebuhr: 184
— *Philosophica*, edited by Erasmus: 77
— *De re publica*, edited by A. Mai: 184
— — 'Dream of Scipio': 6
— *De legibus*, commentary of Turnebus: 112
— translation of Aratus' *Phaenomena*: 126
— Vettori's work on *Epistulae*, *Philosophica*, and *Rhetorica*: 136
— imitation of C.'s language and style, *see* Ciceronianism
— Dolet's exposition of Ciceronian usage: 107
— theory and practice of prose rhythm, discussed by Bruni: 30
— anapaestic fragment of a Latin tragedy discovered by Bentley in a poetical quotation of C.: 150
Ciceronianism as humanistic imitation of Ciceronian language: 30
— and Politian: 43
— founded by G. da Barzizza and Guarino da Verona: 43
— propagated by the schoolmasters, not by the great scholars: 53
— Erasmus's *Ciceronianus*: 79
Ciriaco di Ancona: 51
Clark, A. C.: 32, 40.1
classici, writers of the first class, term coined by Beatus Rhenanus (?): 84

classicism, English, and the 'Society of Dilettanti': 162
'classics', *see classici*
Clement of Alexandria, edition of P. Vettori: 136
— Strom. VI 8.67.1: 70
Cluny: 32
Cobet, C. G.: 163
coinage, Roman, Budé's study of: 101
Colet, John: *72 f.
— and Erasmus: 72 f., 74, 79
Collège de Coqueret: 103 f.
Collège Royal (Collège de France): 102, 111, 120
Collegium trilingue (Louvain): 73, 89, 125
Collins, Antony, *Discourse on Freethinking*, and Bentley: 148, 156–8
Cologne, university, and the controversy about Jewish books: 89
Colonna, Giovanni, cardinal: 40.3, 54
Colonos, K. O. Müller buried on the hill of: 187
Columbus, Christopher, and Enea Silvio Piccolomini: 60
Comes, Natalis: 21
Comma Ioanneum, *see* Bible, NT, I John 5: 7 and 8
Commelinus, printer and publisher in Heidelberg: 141
commentary:
— distinction between running c. and monographs made by Erasmus: 78 f.
— commentaries in the Italian Renaissance: 54 f.
— Casaubon's commentaries: 121
Complutensis (Polyglot Bible of Alcalà, Complutum): 65, 94 f.
Cono (or Kuno), John: 86.9
Constance, council of: 31, 33
Conti, Natale, *see* Comes, Natalis
Copenhagen: 183
Cordova: 94
Cortese, Paolo: 43
Corvey, monastery of: 85.1
cosmopolitanism, of Erasmus: 73
Crete, Greek scriptoria: 56
Creuzer, Friedrich: 187
criticism, biblical: 37 (Valla), 76 (Valla and Erasmus), 128.6 (Spinoza), 130 (R. Simon), 156 (Bentley), 190 (Lachmann)
— historical, born from Niebuhr's work on Livy: 184
— literary: 27 (Salutati), 146 f. (Bentley); *see also* criticism, textual
— textual, revived by Petrarch: 8
— Salutati's attention to problems of: 26
— Poggio's 'emendare': 32

criticism, *contd.*
— started with Valla's work on the text of Livy: 36 ff.
— Valla's biblical criticism: 37 f.
— Valla and the authenticity of the *Donatio Constantini*: 39 f.
— Politian's principles: 44
— Petrarch denied the authenticity of documents sent him by the Emperor Charles IV: 58
— — Erasmus's principles: 74
— — Robortello, *Disputatio de arte critica*, G. J. Vossius, *Aristarchus*, and J. Le Clere, *Ars critica*: 137
— — Bentley's principles: 153 f.
— — Lachmann's method: 190; *see also* interpretation, archetypus, divination, emendatio, recensio
Cuiacius, J.: 101, 117, 132, 174.6
Curtius, Ernst Robert: 146.5
Curtius Rufus, edited by Erasmus: 77
cycle, epic: 180
Cyprian, Saint: 78

Dacier, Anne: 134, 153
Damm, Christian Tobias: 167
Danès, Pierre: 102
Daniel, Pierre: 132
D'Annunzio, Gabriele: 21
Dante, beloved by Boccaccio: 20
— Boccaccio's lectures and commentary on the *Commedia*: 23 f.
Dawes, Richard, *Miscellanea critica*: 157, 161
Decembrio, Pier Candido, tells us the date of Petrarch's death: 14
— epitome of Plutarch: 29
deism, attacked by Bentley: 147
Delos, colossus of the Naxians, inscription: 155
'Demogorgon': 21 f.
Demosthenes, Latin translation of smaller excerpts by Valla: 38.5
— edited by Erasmus: 77
— edited by H. Wolf: 140
— Codex Monacensis Graecus 485: 140
Deventer: 69, 71, 90
Devotio moderna: *69 f.
— Bohemia the link between Italian early humanism and Dev.m. (?): 59.2
— and Erasmus: 71 f.
— and Colet: 72
— influence on Wimpfeling and Sturm: 83
dialectology, advanced to a special branch of classical scholarship by Ahrens's methodical treatment: 188
dialects, Greek, study of (Ahrens): 187 f.; *see also* Hellenistic language

digamma, discovered by Bentley: 157
Digests, Florentine manuscript: 62 f.
— — edited by A. Agustín: 95
Dilettanti, Society of: 161 f.
Dilthey, Wilhelm: 19
dimeter, anapaestic, Bentley's observation: 149
Diodorus Siculus, Poggio's Latin translation of the first five books: 34
Dionigi de' Roberti, Augustinian monk, friend of Petrarch: 11
— commentaries and explanatory notes on Valerius Maximus and a few Roman poets: 54
Dionysius Areopagita, and Valla: 40
Dionysius of Halicarnassus, edited by P. Vettori: 136
Dionysius Thrax, used by Chrysoloras in his Ἐρωτήματα: 53
Dithmarschen: 184
'divination': 154 (Bentley), 161 (Porson)
Dobree, P. P., work on the Attic orators: 161
Döllinger, Ignaz, and Lord Acton: 77
Dolet, Étienne: 107
Dollinger, Heinz: 139.7
Dominicans, and the controversy about Jewish books: 89
Donatio Constantini, Valla's criticism: 39 f.
Donatus, commentary on Terence, known to Salutati and used by Bentley: 155
Dorat, Jean: *102–7
— and J. J. Scaliger: 115
— and W. Canter: 125
— Porson's emendations in the text of Aeschylus comparable in quality to those of D.: 160
Dordrecht, Synod of: 127
Doric, K. O. Müller's hymn on the excellence of everything D.: 187
— dialect (Ahrens, *De dialecto Dorica*): 188
Dousa, Franciscus: 119, 128
Dousa, Janus: 119, 128
Dresden: 168, 171
Droysen, Johann Gustav: 188 f.
Dryden, John: 21 n. 7
Du Bellay, Joachim, and C. de Seyssel: 99.4
Du Cange, Charles: 133
Dukas, Demetrios: 65

Edinburgh: 183
editones principes of Greek and Latin classics: 50 (Italian Renaissance in general), 56 (Aldine Press), 85 (Velleius Paterculus), 106 (*Anacreontea*), 108 (Eusebius, *Hist. eccl.*), 109 (eighteen first editions of H. Étienne), 136 (Scholia to Aeschylus), 140 (Zonaras, N. Choniates, Nikephoros

Gregoras), 141 (Photius, *Bibl.*, Procopius, Phrynichus), 184 (Gaius, *Inst.*, Merobaudes, Cicero, *De re p.*)

editones variorum: 162

education, in the Italian humanism: 52

— classical, and E. S. Piccolomini: 60

— — in England: 159

Egyptian art: 170

Einsiedeln: 31

Elgin Marbles: 162

Elgin, Lord: 162

Elizabeth I, Queen of England: 110, 143

Elmsley, P.: 161, 173

Elzevier, family: 138

— edition of the Greek NT of 1633: 109

'emendatio': 190; cf. also Poggio's 'emendare': 32

encyclopaedia, Wolf's lectures about: 174.6, 175

encyclopedism, in Holland: 162

England:

— Poggio in E.: 33

— humanism during the fifteenth century: 61 f.

— — during the early sixteenth century: 66

— J. J. Scaliger in E.: 116 f.

— Casaubon in E.: 120

— humanism in the time of Colet and of Th. More: 72 f.

— classical scholarship in the seventeenth and eighteenth centuries: 143 ff.

— G. Hermann's objection to the method of the English scholars: 178

— Niebuhr's relation to E.: 183.3

Ennius: 6 f.

Epictetus, *Enchiridion*, version of N. Perotti: 54.6

Epicureanism, and Valla: 40

Epigrammata Bobiensia: 76.6

epigrams, Greek explained by Dorat: 104

Epimenides: 71.5

Epistolographi, termed 'declamatiunculae' by Erasmus: 75 f.

epistolography, humanistic, influenced by Cicero's *Epistulae*: 26

Epistulae obscurorum virorum: 89 f.

Erasmus of Rotterdam: *71–81

— edition of his works by J. Le Clerc: 71.1, 137

— edition of Greek NT: 73, 74, 76 f., 83, 94

— — and the Complutensis: 66

— — Stunica's attacks: 94

— — third edition and the edition of R. Étienne: 108

— edition of Valla's *Adnotationes*: 38

— *Ratio verae theologiae*: 92

— and Giovanni Pontano: 57

— and C. Peutinger: 63

— and Robert Gaguin: 65

— and the humanistic circle on the upper Rhine: 82 ff.

— and B. Rhenanus: 83

— and Zasius: 86

— and Reuchlin: 87

— and Luther: 80, 90 f.

— and Melanchthon: 91

— and Sepulveda: 95, 154.2

— and Lefèvre d'Étaples: 100

— and Budé: 101

— and G. de Brie: 103

— J. C. Scaliger's attacks against him: 114.2

— J. J. Scaliger admired his greatness: 114.2

— Erasmian spirit of H. Grotius' *De iure belli ac pacis*: 127

— Bentley, as true inheritor of Erasmian tradition insisted on the necessity of critical studies in their application to Scripture: 156

— and Spain: 95

— fight against the barbarians, ironical mockery of Alexander of Villedieu: 52

— and Ciceronianism: 54, 79

— and the problem of the pronunciation of Greek: 53, 79, 88

— interest in ancient fables: 91

— revived the genre of 'loci' in his *Ratio verae theologiae*: 92

— on the Epistles of Phalaris: 75, 152.1

— rejected the Comma Ioanneum: 160

Eratosthenes: 101, 122

Erfurt, university: 93

Ernesti, Christian Gottlieb: 172

Ernesti, Johann August: 171 f.

Escorial, library: 95

Étiennes, family, press in Paris: 136

Étienne, Henri I: 107

Étienne, Henri II: *109 f.

— in audience of the *lecteurs royaux* for Greek: 102

— editio princeps of the *Anacreontea*: 104, 106 f., 109

— complete collection of Greek epic poets: 106

— edition of Aeschylus: 106

— edition of Sophocles: 111

— tried to make a collection of the Greek lyric fragments: 109, 153

— and Hieronymus Wolf: 140

— *Poetae Graeci*: 157

— devoted attention to the study of Greek dialects: 188

Étienne, Robert: *107 f.

— folio edition of the Greek NT (1550) and the edition of Erasmus: 77

Eton: 144, 160
Etruria, art of: 170
Etruscans: 187
Etymologicum Magnum, ed. by Gaisford: 153
Etymologies, Greek: 150
Euripides, ed. by W. Canter: 125
— ed. by P. Vettori: 136
— ed. by J. Barnes: 161
— *Hecuba* and *Iphigenia*, translated by Erasmus: 77
— *Phoenissae*, translated into Latin by H. Grotius: 127
— Κρῆτες, and Bentley: 149
— Porson's editions of *Phoe.*, *Hec.*, *Or.*, and *Med.*: 160.3
— Musgrave's *Exercitationes in Euripidem*: 161
Europe, supranational unity: 170
Eusebius, *Chronicle*: 12.1 (Petrarch's copy), 118 (Jerome's translation and J.J. Scaliger)
— *hist. eccl.*, editio princeps: 108
Eve of St. Bartholomew: 114
Evelyn, John: 147

Faber Stapulensis, *see* Lefèvre d'Étaples
Fabri, Anna *see* Dacier, Anne
Fabricius, Johann Andreas: 173
faith alone the way to Christian truth: 92
Feliciano, Felice: 51
feritas, in opposition to *humanitas*: 15
Ferrara, library: 50
— P. Luder in F.: 63
— R. Agricola in F.: 64
— Guarino's school: 66
Festus, Sextus Pompeius, ed. by A. Agustín: 95
— ed. by Scaliger: 115
Ficino, Marsilio: *57
— and John Colet: 72
Fiesole, Benedictine abbey of Badia, library: 49
Filelfo, Francesco: 48 f.
finances, Greek, studied by Böckh: 181
Finariensis, Petrus Antonius: 63.1
Fisher, Christopher: 38
Fisher, John, cardinal: 79
Flacius Illyricus, Matthias: 93
Florence, humanism: 25 ff.
— beginning of Greek studies: 27
— Council of: 37.8, 57
— Politian and the Medicean circle: 42 ff.
— libraries: 49
— Dominican convent of S. Marco: 49
— 'studio fiorentino': 56
Fonseca, Archbishop of Toledo, included St. Augustine among the 'classici': 84.4
Forcellini, *Totius Latinitatis Lexicon*: 107

Fouquet, Jean: 22
Francis I, King of France: 99, 102
François de Sales, Saint: 102, 130
fratres communis vitae see Devotio moderna
Frederick III, Emperor: 59
Frederick II, King of Prussia: 172
Freiburg im Breisgau, and Erasmus: 73
— and Zasius: 86
friendship, desire for, in Petrarch's letters: 10
Froben, John, printer of Erasmus: 66, 73, 78, 83
Frulovisi, Tito Livio: 62
Fugger, family: 140 f.
Fugger, Johann Jacob, library: 140
Fulda: 31
Fulgentius, and Boccaccio: 24
Funus linguae Hellenisticae (anonymous writing, often ascribed to Salmasius): 123.1, 189.4

Gaguin, Robert: 65, 73
Gaisford, Thomas: 153
Gaius, *Institutiones*, palimpsest, discovered by Niebuhr: 184
Gale, Thomas: 21.3, 144
Galen, translated by Erasmus: 77
Galilei, Galileo: 114
Garamond, Claude: 107
Garin, Eugenio: 19
Gasparino, Veronese scholar: 56
Gataker, Thomas: 144
Gelenius, Sigismund: 85
Geneva: 117 (J. J. Scaliger), 120 (Casaubon)
geography, E. S. Piccolomini's works on: 60
German antiquity, interest in, revived by Celtis: 64
Germanic languages: 185
Germany:
— prehumanism: 59
— and Italian humanism: 62 ff.
— humanistic circle on the Upper Rhine: 82 ff.
— foundation of Greek Studies: 86 f.
— classical scholarship in the sixteenth and seventeenth centuries: 86 ff., 139 ff.
— Winckelmann and Neohellenism: 167 ff.
Gerson, Jean Charlier de: 61
Gesner, Conrad: 140.7
Gesner, Johann Matthias: 168, 175
Ghellinck, J. de, on Erasmus's edition of St. Augustine: 78.2
Gibbon, Edward: *162
— and Tillemont: 133
— applauded Porson's proof of the spuriousness of the Comma Ioanneum: 160
Glareanus, *see* Loriti
gnomic poets, Greek, ed. by Turnebus: III
Görres, Joseph: 187

Goethe, J. W. v., and Erasmus about 'venerari': 75
— *Winckelmann und sein Jahrhundert*: 169, 185
— and the Hellenism: 170
— enthusiastic reception of Rob. Wood's *Essay on the original genius of Homer*: 175
— and F. A. Wolf: 176.6, 177
— 'Philologie ohne Kunstbegriff nur einäugig': 180.1
Göttingen: 163, 168, 173, 176, 185, 186, 188
Gogavius: 154.3
Gorgias, hymn on the λόγος: 35.2
'Gothic', as term of abuse in a stylistic context coined by Valla: 35
— script (Gothico-Antiqua) *see* script
Graevius, Johann Georg, edition of Callimachus: 152, 162
grammar, first of the literary arts: 52 f.
— Scaliger's conception of: 116
— German, J. Grimm's *Deutsche Grammatik*: 185
— Greek, the 'Ερωτήματα of M. Chrysoloras and of C. Lascaris: 53
— — elementary of Georg Simler: 88
— — Melanchthon's Greek gr.: 88, 91
— — treated by Dawes in his *Miscellanea critica*: 161
— — Buttmann's *Greek Grammar*: 185
— — R. Kühner's manual: 188
— *see also grammatica* and γραμματικός, grammaticus
grammarians, Greek: 155
grammatica and *religio*: 116 (Scaliger)
grammaticus:
— Politian never claimed to be a philosopher, but only a 'gr.': 58
— Scaliger: 'Utinam essem bonus grammaticus': 116
Gratius, Ortwin: 89
Gray, Lady Jane: 143
Greban, Arnoul: 21
Grebenstein, Heinrich von: 31.1
Greece, and K. O. Müller: 186
Greek:
— a dream-world in Petrarch's time: 13
— Poggio learning some Greek: 34
— W. Grocyn the first to teach Greek in an English university: 66
— taught in Paris: 72.1
— importance of its knowledge: 53 (B. Guarino), 69 (Hegius), 73 (Colet and Scaliger)
— problem of the pronunciation: 53, 79, 88
— love of, in the age of Winckelmann: 167 ff.
— Welcker's knowledge and understanding of Greek: 180
— *see also* grammar, Greek *and* Greek studies

Greek studies:
— beginning in Florence: 27
— no Greek manuscripts detected in Salutati's library: 27
— Bruni's translations of Greek literature into Latin: 27 ff.
— of Politian: 43 ff.
— the beginnings in Germany: 86 ff.
— in Wittenberg: 91 f., 93 f.
— in the French Renaissance: 101 (Budé), 103 (Dorat), 108 f. (H. Étienne), 115 (J. J. Scaliger)
— of W. Canter: 125 f.
— in England: 159 ff.
— 'tyranny' of Greece over Germany (?): 168.5
— the superiority of the creations of the Greeks demonstrated by F. Schlegel: 186
Greene, Robert: 21.7
Gregoras, Nikephoros, first ed. by H. Wolf: 140
Gregorius Naz., well known by Erasmus: 73.8
Grimm, Jacob: *185
— influence on Ahrens: 188
Griselda, story of, translated into Latin by Petrarch: 20
Grocyn, William: 66
Gronovius, Jacob: 129
Gronovius, Johann Friedrich: 129
Groot, Hugo de, *see* Grotius, Hugo
Groote, Geert: 69
Grotius, Hugo: *126–8
— and Scaliger: 119, 126
— on 'nescire': 75
— and Euripides' Κρῆτες: 149
— read by Winckelmann: 168
Grynaeus, Simon: 85, 139
Guarino da Verona, recites passages of Poggio's Latin letters in Valla's 'apologus': 36
— and the foundation of Ciceronianism: 43
— *Regulae grammaticae*: 52
— translated excerpts from Chrysoloras's 'Ερωτήματα into Latin: 53
Guarino, Battista, *De ordine docendi et studendi*: 53
— taught Aldus Manutius Greek and Latin: 56
— his school at Ferrara: 63, 66
Gustavus Adolphus, King of Sweden: 128
gymnasium, humanistic:, 170
Gyraldus, L. G.: 21.3

Hainrode: 173
Hales, John: 144
Halle: 168, 176, 181, 186

Halm, Karl: 107.4
Hamburg: 163
Hannover: 188
Hardouin, Jean: 135
harmony of classical poetry, to be restored by the true critic (Bentley): 154; cf. also 147 and 158
Headlam, Walter: 112
Heath, Benjamin: 161
Hebraism (καρδίᾳ πλανώμενοι), detected by Bentley in a spurious quotation from Sophocles: 150
Hebrew studies in Germany, founded by Reuchlin: 86 f.
Hegel, G. F. W.: 171, 189
Hegius, Alexander: 69
Heidelberg: 63, 122, 138, 140 f.
Heinsius, Daniel: 119, 129
Heinsius, Nicolaus: 129
Hellenism, prepared the way for Christianity (Budé): 102
'Hellenismus', term for the hellenistic age: 189
Hellenistic age: 123, 189; *see also* 'Hellenismus'.
Hellenistic language (*lingua Hellenistica*): 123, 189
Hemsterhuys, Tiberius: 163
— edition of Pollux with Bentley's help: 153
Henninius, H. C.: 89.3
Henry IV, King of France: 117, 120
Herculaneum, excavations, and Winckelmann: 170
Herder, J. G., and Winckelmann: 169
— *Kritische Wälder*: 170
— the humanistic approach fruitful in his writings: 171
— and the ideas of *Ossian* and Percy's *Relics*: 175
Hermann, Gottfried: *178 f.
— on 'nescire': 75
— on Dorat's importance: 103.3; cf. also 102.4
— and Welcker: 180
— reconstructions of lost Greek tragedies: 180
— and Böckh: 181
— Lachmann called him his 'pater studiorum': 190
hermeneutic: 92 f., 181
Herodotus, translated into Latin by Valla: 38
— H. Étienne's *Traité préparatif à l'apologie pour Hérodote*: 110
Hersfeld: 85
Hesiod, *Theogony*, proem: 6
— — Kronos: 22

— *Op.* 293 ff., cited by Petrarch: 8
— Politian's lectures on H.: 44
— G. Hermann's analysis of the text of H.: 179
Hesychius: 150, 163
hexameter, epic, history of, given by G. Hermann in his *Orphica*: 179
Heyne, Christian Gottlob: *171, 173, 181
— on 'demogorgon': 22.3
Hilary, Saint: 78
historia: 174 ('reverenda est historia', F. A. Wolf): 186
historians, German, of the nineteenth century: 183.2, 188
— Greek, translation into Latin commissioned by pope Nicolas V: 38
— — treated by Vossius: 137
— Roman, interest in France: 61
historicism: 190
historiography of the Italian Renaissance: 30 (Bruni), 51 (Biondo), 60 (E. S. Piccolomini)
history, B. Rhenanus's conception of: 84
— F. A. Wolf's reverence for: 174, 186
— Böckh's lively sense of: 181
— Droysen's concept of: 188.2
— *see also* criticism, historical
— ecclesiastical, of the later Roman empire: 134
Hobbes, Thomas: 146
Hody, Humphrey: 149
Höschel, David: *141
— and Georgios Synkellos: 118.3
Holbein, Hans, the younger, pictures of Erasmus: 76
Holland, humanism of the Devotio moderna: 69 f.
— classical scholarship: 124 ff., 137 ff., 162 f.
Holtzmann, Wilhelm (Xylander): 140 f.
Homer, his own interpreter: 76
— Peisistratean recension: 158
— 'Alexandrian' text, tried to restore by F. A. Wolf: 174
— F. A. Wolf and the problem of the origin and the unity of Homeric poems: 174 ff.
— 'Omero poeta sovrano' (Dante): 24
— in Erasmus's time no longer the centre of scholarship, but the New Testament: 76
— 'la naïve facilité d'Homère' (Ronsard): 103.3
— and Petrarch: 6, 9.5, 14
— and Boccaccio: 24
— and Politian: 44
— and Melanchthon: 91
— and Dorat: 104, 106
— allegorical interpretation of Dorat and his contemporaries: 104.3

— French translations of the sixteenth and seventeenth centuries: 134
— and d'Aubignac: 154
— Pope's translations of the *Il.* and the *Od.* and Bentley's judgement: 157 f.
— edition prepared by Bentley: 156 f.
— Bentley's judgement on Homeric poetry: 158
— R. Wood, *Essay on the original genius and writings of Homer*: 161, 175
— Ch. T. Damm (described as 'Ὁμηρικώτατος), etymological Homeric dictionary and a translation of the *Iliad* and the *Odyssey* into German prose: 168
— and F. A. Wolf: 173–5
— edition of I. Bekker: 181
— G. Hermann's analysis of the text of *Il.* and *Od.*: 179
— *Iliad*, Codex Ven. Marc. 454: 48, 174
— — books II–V translated into Latin by Politian: 43
— — IX 222–603, translated into Latin prose by Bruni: 28
— — translated into Latin by Valla: 38
— — Scholia Victoriana: 136
— — Scholia ed. by Villoison from the Cod. Ven. Marc. 454: 174
— — edition of Turnebus: 111
— *Odyssey*, translated into Latin prose by an anonymous writer before 1398: 14.9
[—] Homeric Hymns, treated by G. Hermann: 179
— Buttmann's *Lexilogus*, one of the most influential books on Homeric language: 185
Homeridae: 174
'homo omnium horarum': 79 (on Th. More), 86 (applied by Erasmus to Glareanus)
'homo politicus' (Lipsius, Grotius): 126
Horace, *c.* I 23.5, Bentley's conjecture: 154
— P. Luder's lectures on H. in the university of Ingolstadt: 63
— Petrus Tritonius made tunes for nineteen odes: 63 f.
— edition of D. Lambinus: 112
— edition of Hardouin: 135.3
— edition of Bentley: 153 f.
Hrosvitha of Gandersheim: 64
Huet, Pierre Daniel: 135
humanism:
— the term 'humanism', 'Humanismus': 17
— the 'dawn' of h.: 4.1 and 7
— Christian: 129
— — in England: 143 f. (Christian-Platonic h.)
— — *see also* philosophia Christi
— Benedictine: 131

— Protestant, of Melanchthon: 91 f.
— 'humanism, devout': 129 f.
— *humanitas*, conception of: 15–17
— and Christian piety (Erasmus, Colet): 72
— of Erasmus, connected with his idea of 'tranquillitas': 80
— *studia humanitatis*: 16, 63
Humboldt, Caroline von: 179.1
Humboldt, Wilhelm von: *184 f.
— theorist of the Neohellenism: 170
— founded the university of Berlin: 176, 183
— and F. A. Wolf: 177
— translation of Aeschylus' *Agamemnon*: 179
— and G. Hermann: 179
— and Welcker: 179, 184.4
— and the Greeks: 181
— J. Grimm to be regarded as one of Humboldt's school: 185
— idea of the state, admired by Mommsen: 190
Humphrey, Duke of Gloucester: 62
Hutten, Ulrich von, edition of Valla's *Declamatio*: 40
— and the *Epistulae obscurorum virorum*: 90
Hyma, A., *editio princeps* of Erasmus's *Antibarbari*: 80.1

ictus, in classical verse, problem of, and Bentley's solution: 155
idealism, post-Kantian, disliked by G. Hermann: 79
Ignatius of Loyola, could not feel sympathy with Erasmus's 'philosophia Christi': 81.2
imitation (Nachahmung), problem of, and Winckelmann: 169 f.
'Indogermanische Sprachwissenschaft', founded by Bopp: 185
Ingolstadt: 63, 87
Inquisition: 96
inscriptions, ancient:
— copied by Ciriaco di Ancona: 51
— studied by A. Agustín: 95
— Janus Gruter's collection: 116, 138
— Bentley's contribution to the text of the inscription on the Naxian colossus at Delos: 155
— Böckh's studies: 181
interpres, regius: 103 f. (Dorat)
— bonus (Erasmian formula): 128
interpretation, method of : 74 (Erasmus)
— Luther's conception of: 90 f.
Ion of Chios: 150
Irenaeus, Saint: 78
Isocrates, ed. by Hieronymus Wolf: 140

Jacopo Antiquario: 42
Jamyn, Amadis: 134

Jena: 168, 176, 188

Jerome, Saint, and Petrarch: 12

— principles of translating (*Epist.* 106): 29.7

— Valla compared his Latin version of the NT with the Greek original: 37

— and Erasmus: 73, 78

— translation of Eusebius' *Chronicle*: 118

Jerome of Prag, Poggio's account of his trial and execution: 33

Jesuits, and Erasmus: 81.2

— , French, of the seventeenth century: 132 f.

Jewish books, should be confiscated and destroyed: 89

Johannes von Neumarkt: 59

John II, the Good, King of France: 60

John Chrysostom, ed. by Erasmus:78

— ed. by H. Savile: 144

John XXIII, pope, praising Latin the 'materna vox': 77

Julius Africanus: 118

jurisprudence, and Erasmian ideas: 86; *see also* law, Roman

Justi, Carl: 168.5

Justinus, ed. by J. Bongars: 132

Juvenal, favourite poet of G. Tortelli: 55

Kant, I.: 171

— and G. Hermann: 178

Kepler, Johannes: 114

Kiel: 183, 188

Kircher, Athanasius: 63.5

Knöringen, Johann Egolph von: 85.4

Kuno, John, *see* Cono

Kristeller, Paul Oskar: 19

Kronos, in Hesiod's *Theogony*: 22

'Küchenlatein' (Latinum culinarium): 36, 90

Kühner, Raphael: 188

Küster, Ludolf, edition of Suidas: 153, 162

Kumarbi, Hurrian god: 21 f.

Kunze, Emil: 171.1

Lachmann, Karl: *190

— the 'textus receptus' of the Greek NT and L.'s critical recension: 109, 156, 190

— Ahrens's *De dialecto Dorica* dedicated to him: 188

— editions of Latin texts: 190

Lactantius Placidus: 22, 24

Lambinus, Dionysius: *112, 135

Landriani, Gerardus, discovery of the triad of Cicero's rhetorical writings: 33

Langres: 32

language, Lachmann the greatest master of (called by Mommsen): 190

— study of, by Humboldt and Bopp: 185;

see also linguistics *and* 'Indogermanische Sprachwissenschaft'

language, French, superior to the other modern languages (H. Étienne): 110

language, Greek, *see* Greek, *and* grammar, Greek

language, Latin, *see* Latin; lexicography, Latin; *Thesaurus Linguae Latinae*

Lascaris, Constantinus: 53

Lascaris, Janus, Lorenzo de' Medici's principal agent in the East: 48

— the problem of the pronunciation of Greek: 53, 89

— in France: 61, 99.4, 102

— invited to the Collegium Trilingue of Louvain: 89

Latimer, William: 66

Latin:

— Petrarch's enthusiasm for the 'sweetness and sonority' of ancient L.: 5, 13

— imitation of Ciceronian language and style, *see* Ciceronianism

— becomes a dead language: 30, 36

— Poggio's mastery of L. as a living language: 33 f.

— Valla's *Elegantiae* and his praise of the Latin language: 35

— 'Latinum culinarium': 36, 90

— Valla and the Latin style of the Church Fathers: 37

— — , analysis of the Latin language of the *Donatio Constantini*: 39

— Pomponio Leto strove for perfection in L.: 51

— kept the dominant position in the Italian Renaissance: 53

— pope John XXIII praising L. the 'materna vox': 77

— J. J. Scaliger, use of L.: 114

— — knowledge of archaic L.: 115, 117

— study of, prevailing in Holland: 125

— verses of H. Grotius: 126

— studies of D. and N. Heinsius: 129

— Bembo's virtuosity in his L. works: 136

— poems of G. Buchanan and others: 144

— break with the L. tradition in Germany in the age of Neohellenism: 170

— F. A. Wolf's Latin 174

— Wilamowitz's Latin studies: 178.2

— Niebuhr's discoveries of Latin texts: 183

— in the course of the nineteenth century L. regained the place beside Greek that it had occupied before Winckelmann: 190

law, international, and H. Grotius: 127

Lebrija, Antonio de: 65 f., 95

Le Clerc, Jean: *137

— edition of Menander and Philemon: 153

— edition of Erasmus's works: 71.1, 137
Lefèvre, Tanneguy (Tanaquil Faber): 135
Lefèvre d'Étaples (Faber Stapulensis): *61, 100
legal studies, *see* law, study of
Leibniz, G. W.: 158
Leipzig: 163, 168, 171, 172, 176
Lenain de Tillemont, Sébastien: 133
Leo X, pope: 48, 52 f., 66, 77, 89
Lessing, G. E.: 170–2
Leto, Pomponio: *51 f.
— commentaries on the whole of Virgil (?): 55
— Celtis his pupil: 63
Lewis, G. C.: 183.3
lexicographers, Greek: 150, 153, 163
lexicography, Greek: 101 (Budé, *Commentarii Linguae Graecae*), 110 (H. Étienne's *Thesaurus Graecae Linguae*), 133 (Du Cange)
lexicography, Latin: 107 (R. Étienne, É. Dolet, Forcellini), 133 (Du Cange)
Leyden, university: 117, 122 f., 124, 129
libraries, great new of the Italian Renaissance: 47, *49 f.
— library of the Kings of Aragon in Naples: 49, 56.7
— l. of the Escorial: 95
— Oxford, Bodleian Library: 119.7, 120.1, 145, 149
— Berne library and the Bongars collection: 132
— Paris, Bibliothèque nationale, and Petrarch's books: 13.1
— — Petrarch's autographs and marginal notes among the manuscripts: 5.1
— — and Huet's books: 135
— Heidelberg, Palatine library: 128
— Verona, Capitular Library: 184
— private: 12 f. (Petrarch), 26 (Salutati), 30 f. (Niccoli), 44.7 and 45 (Politian), 49 (Bessarion), 87 (Reuchlin), 95 (Agustín), 132 (Bongars, P. Daniel, Cuiacius), 135 (Huet), 139.7 (H. Wolf), 140 (Fugger family), 145 (Dr. Stillingfleet), 168 (count H. v. Bünau), 171 (count H. v. Brühl)
Ligurinus, discovered by Celtis: 64
Linacre, Thomas: 66
lingua Hellenistica, *see* Hellenistic language
linguistics: 185 (Humboldt, Bopp, J. Grimm), 188 (Ahrens)
Lipsius, Justus: *124–6, 132
literae and *humanitas*: 15
literature, Greek, history of: 187.3 (K. O. Müller, Bernhardy)
Livy, the 'Codex Regius': 37
— books 41–5 discovered by S. Grynaeus in 1527: 85

— ed. by Erasmus: 77
— ed. by B. Rhenanus and S. Gelenius (1535): 85
— and Petrarch: 6 ff., 36
— and Boccaccio: 23
— Valla's *Emendationes Livianae*: 8, 36 f., 184
— translated into French by P. Bersuire: 60
— his books divided into chapters by Gruter: 138
— Gruter's work on him: 138
— Niebuhr's criticism: 184
loci: 92
Locke, John: 147
London: 145–7; *see also* Museum, British
London Society for the Diffusion of Useful Knowledge: 187
[Longinus] Περὶ ὕψους, ed. by Robortello: 137
Loriti, Heinrich (Glareanus): *85, 137
Lorsch, monastery of: 85
Louis XII, King of France: 102
Louvain, Collegium Trilingue: 73, 89, 96, 125
Lovati, Lovato: 4
Lucan, and Politian: 43
— ed. by H. Grotius: 127
Lucian: 77
Lucretius, ms. discovered by Poggio: 33
— treated by Giovanni Pontano: 57.1
— Lucretian scholarship in the Quattrocento: 57.1
— Lambinus's edition: 112
— ed. by Lachmann: 190
Luder, Peter: 63
Luther, M.: *90 f.
— lecture on the Epistle to the Romans and the German translation of the NT and Erasmus's edition of the Greek NT: 77
— opponent of Erasmian humanism: 80
— interpreting N.T. Rom. 1:17: 90
— and 'pagan' Greek and Roman literature: 90 f.
— and Melanchthon: 91 f.
Lycophron, and Dorat: 104
— translated into Latin by J. J. Scaliger: 115
— a Callimachean line in the Scholia: 153.4
Lydus, Johannes, used by Politian: 45.6

Maas, Paul: 188.1, 190.1
Mabillon, Jean: 131
Macpherson, James, *Ossian*: 175
Macrobius: 14
Mai, Angelo, edition of Cicero *De re publica* from a Vatican palimpsest: 184
Malalas, Johannes: 149
'man of all seasons' (omnium horarum homo): 79, 86

Manilius, rediscovered by Poggio: 33
— first printed edition of 1538: 37.8
— J. J. Scaliger's edition of 1579: 117
— Bentley's edition: 153, 155
manuscripts:
— Oxford, Exeter College 186 (Petrarch's Suetonius): 13.9
— Bavarian State Library, Munich, cod. gall. 6 (French translation of Boccacio *De casibus*): 22
— Codex Casinensis of Tacitus: 23
— Bruni collecting and copying mss.: 30
— Poggio as hunter of mss.: 31
— Codex Fuldensis of Tacitus: 31
— Cod. Vat. lat. 11458 as specimen of Poggio's handwriting: 32
— the 'Codex Regius' of Livy: 37
— Codex Pisanus of the *Pandects*: 43.5
— Politian the first to make complete collations of mss.: 44
— travellers between West and East in quest of Greek mss.: 48
— Codex Venet. Marc. 454 of the *Iliad*: 48, 174
— Codex Laur. L 10 (Varro, *De L.L.*): 24
— Codex Laur. XXXII 2 (Euripides): 14.5
— Codex Laur. XXXII 9 (Aeschylus, Sophocles, Apollonius Rhodius): 48, 111, 136
— Codex Palatinus 23 (*Greek Anthology*): 48 122, 138
— collection of Humphrey, Duke of Gloucester: 62
— the Florentine ms. of the *Digests*: 62, 95
— Greek, used by Erasmus for his edition of the NT: 76 f.
— of Livy's fourth and fifth decades from Murbach, Lorsch, Worms, and Speyer: 85
— of Ammianus Marcellinus from Hersfeld, lost: 85
— Codex Mediceus I of Tacitus: 85.1
— Clm 28325 (Glareanus's poem on the battle of Näfels): 85.4
— Roman legal, discovered by Zasius: 86
— Codex Vat. gr. 1209 (Vat. B) of the Bible: 95
— Codex Parisin. gr. 2711: 111.8
— Cambridge ms. of Origen's *Contra Celsum* and J. J. Scaliger: 116
— Codex Bern. 711 (Ovid, *Ibis*): 132
— Florentine mss. of several of Sophocles' plays collated by Vettori: 136
— Heidelberg, Palatine mss., presented by Maximilian of Bavaria to the Vatican library: 138
— Codices Palatini B and C and Codex Vat. D of Plautus: 139

— Codex Augustanus (A, now Monacensis gr. 485) of Demosthenes: 140
— Codex Blandinius vetustissimus of Horace, highly esteemed by Bentley: 154
— Vatican ms. with fragments of some speeches of Cicero, found by Niebuhr: 184
— Vatican palimpsest containing Cicero *De re publica*: 184
Manutius, Aldus: *56, 66
Marcus Aurelius, *see* Antoninus, Marcus Aurelius
Margaret of Navarra: 61
Markland, Jeremiah: 161
Marlborough, Earl of: 147
Marlowe, Christopher: 21.7
Marmor Parium: 119
Martial, Boccaccio the first to get hold of: 24
— N. Perotti's commentary (*Cornucopiae*): 54
— IX 76.9 ff.: 76.6
Martianus Capella, commentary of H. Grotius: 126
Maurists: 130 f.
Maximilian I of Bavaria: 138
Maximilian II, Emperor: 110
Mead, Richard, editor of Nicander, *Ther.*: 155
Medici, Cosimo de', presented King Alfonso of Aragon with the 'Codex Regius' of Livy: 36 f.
— palace library: 49
— deeply impressed by the Neoplatonism of Plethon: 57
Medici, Lorenzo de', lamented by Politian in a Latin ode: 42
— the first centuria of Politian's *Miscellanea* dedicated to him: 45
— Janus Lascaris his principal agent in the East: 48
— palace library: 49
Melanchthon, Philipp: *91-4
— and Reuchlin: 88
— pupils: 139
Menander, ed. by J. LeClerc: 153
Mendelssohn-Bartholdy, Felix, and J. G. Droysen: 188 f.
Meredith, George: 21.7
Merobaudes, first edition by Niebuhr: 184
Messene: 152
metre:
— metrical system of Latin dramatists: 155
— Bentley's observation of the anapaestic dimeter: 149
— Bentley's 'De metris Terentianis σχεδίασμα': 154
— 'Porson's Law': 160
— history of the epic hexameter given by G. Hermann in his *Orphica*: 179
— Böckh's studies: 181

— *see also* prose rhythm
metrical compositions: 63 f. (Celtis, Petrus Tritonius), 105 (Pléiade)
metrology, Roman, study of: 101 (Budé), 181 (Böckh)
Metsys, Quentin, portrait of Erasmus: 76
Michelet, Jules, and the term 'Renaissance': 18
Middle Ages and Renaissance: 18 f.
'Middle Ages', 'media antiquitas': 84
Milan, libraries: 49 f.
miles Christianus (Erasmus): 81
Mill, John, and Bentley: 145, 149
— edition of the Greek New Testament: 156
Milton, *Paradise lost* II 965 ('Demogorgon'): 21.8
— — ed. by Bentley: 155.5, 157
— *Pro populo Anglicano Defensio*: 122.4
Minervius, Simon: 63.5
Minos: 150
minuscule, Carolingian, and humanistic script: 13
'monachus Hersfeldensis' *see* Grebenstein, Heinrich von
Mommsen, Theodor: *190
— acknowledged Felice Feliciano's merits: 51
— admired Agustín's work on the *Digests*: 95
— acknowledged the accuracy of Bongars: 132
Monk, J. H.: 148.8, 161
Montaigne, Michel de, and E. Pasquier: 105
— and Muret: 113
— and Amyot's French Plutarch: 113
Monte Cassino, monastery, library, visited by Boccaccio: 23
Monte, Piero de: 62
Montfaucon, Bernard de: 131 f.
— and the inscription on the Naxian colossus at Delos: 155.7
Montpellier: 120
Montreuil, Jean de: 61, 64
monuments, ancient, interest in: 50 f.; *see also* antiquities
Morandi, Benedetto: 41.1
More, Thomas, translated into English a 'Life of Pico': 58
— attended Th. Linacre's lectures: 66
— *Utopia*: 79 f.
— and Erasmus: 73, 74.4, 79 ('omnium horarum homo')
— Bremond's biography: 130.1
Moretus, printer in Antwerp: 138; *see also* Plantin-Moretus Museum
Mountjoy, Lord: 72
Müller, Karl Otfried: *186 f.

— the influence of the studies of Roman antiquities in Holland on him: 125.3
— and H. L. Ahrens: 188
Munich, Bavarian State Library, Cod. gall. 6 (French translation of Boccaccio's *De casibus*, illuminated by Jean Fouquet and his pupils): 22
— P. Vettori's books and manuscripts: 136
— the 'Camerariana': 139
— Codex Mon. graec. 485 (Demosthenes): 140
— the library of J. J. Fugger bought by Duke Albrecht V of Bavaria: 140
Murbach, monastery of: 85 f.
Muret, Marc-Antoine de: *112 f.
— commentary on Ronsard's *Amours*: 106
— and J. J. Scaliger: 116
— works in the Latin field: 135
Muses, call a poet: 6
Museum, Alexandrian, and the Collège Royal: 102
Museum, British: 136, 162
Musgrave, Samuel: 161 f.
music, and poetry, *see* metrical compositions
Mussato, Albertino: 4
Musurus, Marcus, edited Plato for the Aldine press: 57
— Celtis his pupil in Padua: 63
— Cono attended his lectures in Padua: 86.9
— living and teaching in Venice: 88
Mutianus Rufus: 90
mysticism, Spanish, influenced by Neoplatonism: 58
Mythographus Vaticanus III: 20
mythology, ancient, in the Italian Renaissance: 20 f.
— — Welcker's work on: 179 f.
— — K. O. Müller's historical theory of mythology and the opposition expressed by F. Creuzer: 186 f.

Näfels, battle at: 85.1
Naples, learned circle of King Alfonso of Aragon: 35 f., 56
— libraries: 49, 56.7
— and Winckelmann: 170
Napoleonic Wars: 176
native tongue, used by Bentley: 151 and by F. A. Wolf: 174
Naucellius: 76.6
Nebrija, Antonio, *see* Lebrija, Antonio de
Neohellenism, German: *165 ff., 183.1 184, 190
Neoplatonism *see* Platonism
Netherlands, Spanish: 96
Newton, Isaac, and Bentley: 144–7, 152
Nicaea, council of: 156

Nicander: 155
Nicholas of Cusa, examination of the *Donatio Constantini*: 39
— and Valla: 40.1
— link between Italy and Germany: 64
— educated in a Deventer school: 69
— as collector of classical manuscripts: 139
Niccoli, Niccolò: *30 f.
— about Petrarch and *studia humanitatis*: 16
— and the Codex Casinensis of Tacitus: 23
— Buondelmonti's *Descriptio Candiae* dedicated to him: 29
Nicolas V, pope, commissioned translations of Greek historians into Latin: 38
— and the Vatican library: 49
Niebuhr, B. G.: *183 f.
— not the first to appreciate the greatness of Scaliger: 170.2
— had no genuine appreciation of great poetry: 188
Niebuhr, Karsten: 183
Niethammer, F., coined the term 'Humanismus' for an educational theory: 17
Nolhac, Pierre de, and Petrarch: 5.1
Nonnus, *Dionysiaca*, read and explained by Dorat: 104
Nuñez, Hernán: 65
Nordhausen: 173
Nuremberg: 62, 86.9, 88, 94, 139

Oeser, Friedrich: 167.4, 169
Olympia, German excavations: 171.1
opera, origin, and Politian's *Orfeo*: 43
Orators, Attic, ed. by H. Wolf and Reiske: 140
— — treated by Dobree: 161
oriental languages, studied by Scaliger: 115; *see also* Arabic studies, Hebrew studies
Origen, and Erasmus: 78
— *Contra Celsum*, Cambridge manuscript and J. J. Scaliger: 116
originality, modern cult of, and imitation (Winckelmann): 170
Orlando di Lasso: 105
Orosius, *Hist.* IV, 19.6: 7
Orphic hymns, translated into Latin by J. J. Scaliger: 115
Orsini, cardinal: 139.5
Ossian *see* Macpherson
Ovid, *Trist.* I 7.7 ff.: 76.6
— — IV 10.2 and Petrarch 'To Posterity': 10 f.
— *Ibis*, and Boccaccio: 24
— — cod. Bernensis 711: 132
— D. Heinsius's criticism: 129
Oxford, university, Duke Humphrey's library: 62

— Bodleian Library: 119.7, 120.1, 145, 149
— colleges: Christ Church: 150
— — St. Edmund's Hall: 145
— — Merton: 144
— — Wadham: 145

Padua: 63
Paestum: 170
'paganism' of the Italian Renaissance: 19
palaeography, Greek, founded by Montfaucon: 131
Palladio: 163.4
Pandects, codex Pisanus, and Politian: 43.5
— Politian's references to the *P.*: 45
— Budé's *Annotationes ad Pandectas*: 101
papacy, attacked by Valla and Hutten: 40
papyri, recently published tragic, and Hermann's and Welcker's reconstructions: 180
paraphrases: 74
Parc, Praemonstratensian abbey: 38
Parentucelli, Tommaso *see* Nicolas V, pope
Paroemiographi Graeci, ed. by Th. Gaisford: 153
Paschalius, Carolus: 126
Pasiphae story, as subject of Euripides' Κρῆτες: 149
Pasquier, Étienne: 105
Pattison, Mark, *Scaliger*: 119.7
— *Isaac Casaubonus*: 120.1
— *F. A. Wolf*: 176.6
Paul II, pope: 52
Paul, Saint, Valla and the authenticity of his correspondence with Seneca: 40
— Pico's lectures on St. P.: 57 f.
— Colet's lectures on the Epistles to the Romans and to the Corinthians: 72
— Luther's lecture on the Epistle to the Romans: 90
Paulinus Venetus, excerpted Tacitus, *Ann.* XIII–XV: 23.3
Pausanias, ed. by Xylander and Sylburg: 141
Pavia: 49, 64
Pearson, John: 144
Peiresc, Claude Favre: 133
Percy, Thomas, *Relics of Ancient English Poetry*: 175
Pergamum: 50
— quarrel between Alexandrians and Pergamenes: 181.1
Perizonius, Jacobus: 163
Perotti, Niccolò, *Rudimenta grammatica*: 52
— *Cornucopiae*: 54 f.
— version of the *Enchiridion* of Epictetus: 54.6
Perrault, Charles: 134
Persian art: 170

— language: 185
Persius, *Satirae*, commentary of Casaubon: 121
Petau, Denys, *Opus de doctrina temporum*: 118
— edition of Synesius (1612): 132
Petau, Paul: 132
Petrarch: *3–16
— at the court of Emperor Charles IV in Prague: 58
— and France: 60
Petronius: 33
Peutinger, Conrad: 62 f.; *see also Tabula Peutingeriana*
Pfefferkorn, Johannes: 89
Pforzheim: 87
Phalaris, Epistles, and Erasmus: 75
— — and Bentley: 147, 149, 150–2
Philemon, ed. by J. Le Clerc: 153
philhellenic society in Venice: 56, 102
Philip II, King of Spain: 95
Philitas, as scholar poet near to Politian: 43
philologia: 101 (Budé), 129 (G. J. Vossius)
— 'studiosus philologiae' (F. A. Wolf): 173
— *perennis*: 176, 180; *see also* 'Altertumswissenschaft'
philology, sacred and profane: 130
philosophia Christi, term coined by R. Agricola: 70 f.
— Erasmus's conception: 74
— — and his thinking about justice and law: 86
— — and Spanish humanism: 95 f.
philosophy, and scholarship: 58 (Politian)
— Greek, studied by Turnebus: 112
— idealistic, and Böckh: 181
— Scholastic, and Erasmus: 71.5, 74 f.
— — and Melanchthon: 92
Phoenician art: 170
Photius, *Lexicon*: 117
— *Bibliotheca*, first edited by D. Höschel: 141
— — ed. by Porson: 153
Phrynichus, first edition of D. Höschel: 141
Piccolomini, Enea Silvio, *see* Pius II, pope
Pico della Mirandola, Giovanni: 57
Pietro da Muglio: 25
piety, Protestant, and new oratory modelled on Cicero in the school of J. Sturm: 84
Pilato, Leonzio, Latin translation of Homer: 14, 21, 24
— and Boccaccio: 24
— not a teacher of Salutati: 27
Pindar, A. Kircher's composition of a melody to the first *Pythian*: 63.5
— and Dorat: 104, 106
— and the Pléiade: 105
— and Böckh: 181
Pirckheimer, Willibald: 62

Pisistratus: 158
Pius II, pope (Enea Silvio Piccolomini): *59 f.
— Voigt's monograph about him: 17
— visit to England and Scotland: 62
Plantin, Christopher: 138
Plantin-Moretus Museum of Printing, Antwerp: 138
Plato, standard edition of H. Étienne: 109
— a new text projected by F. A. Wolf: 173
— ed. by I. Bekker: 181
— *Phaedrus*, recension of Bentley: 153
— *Symposium*, ed. by F. A. Wolf: 173
— six dialogues and some of the *Letters* translated into Latin by Bruni: 29
— Ficino's Latin translation of the dialogues: 57
— É. Dolet the first to translate Platonic dialogues into French: 107
— Schleiermacher's translation into German: 186
— and Petrarch: 14
— birthday, celebrated by the Florentine Platonists: 57
— accepted as 'princeps philosophorum' by the Renaissance: 58
— Dorat introduced his pupils to three poetical dialogues: 104
— F. A. Wolf's favourite author: 173, 186
— in England: 143 f.
— in Germany in the age of Neohellenism: 167
Platonism (including Neoplatonism) of the Italian Renaissance 57 f.
— Florentine: 57 f.
— — and Colet: 72
— — and Reuchlin: 87
— — and Lefèvre d'Étaples: 61
— of Rudolf Agricola: 70
— Christian, not able to compete with Stoicism: 126
— in England: 143 f.
Plautus, copies brought to France by Jean de Montreuil: 65
— and Turnebus, and Muret: 112
— ed. by J. Camerarius: 139
— codd. Palatini B and C, and Vat. D: 139
— and Nicholas of Cusa: 139
Pléiade: 102 f.
Plethon, Georgios Gemistos: 57
Plinius maior, *n.h.*, commentary of Beroaldo: 55.5
— — ed. by Erasmus: 77
Plinius minor, *Ep. ad Trai.*, found in 1500 in Paris: 85.1
Plotinus, translated into Latin by Ficino: 57
— edition of F. Creuzer: 186

Plutarch, Bruni's Latin translation: 29
— and Salutati: 29
— humanistic epitomes: 29.4
— translated by Erasmus: 77
— Amyot's French translation: 113
— and Shakespeare: 113
— edition of Xylander: 140 f.
poem, classical, in analogy to the human body (Bentley): 146 f.
poet:
— *poeta doctus*, scholar poet: 3 (Alexandrians), 42 f., 57 (Beccadelli, G. Pontano)
— ποιητὴς ἅμα καὶ κριτικός: 103 f. (Dorat)
— G. Hermann's belief that the scholar knew what the poet ought to have said: 178
poetic theory of Salutati: 27
— of J. C. Scaliger: 114
poetry, and scholarship, unity of: 3 ff., 8 (Petrarch)
poetry, Greek lyric, studied by Welcker: 180
poetry, Greek, four ages of (Scaliger): 119
poetry, Hellenistic: 150
Poggio: *31–34
— textual criticism: 32, 37
— visit to England: 33, 61 f.
— and Tortelli: 55.3
— interested in ancient fables: 91
Polentonus, Sicco: 24.4
Politian: *42–46
— poetic theory: 27
— and Beroaldo: 55
— and the 'Studio fiorentino': 56
— and Aldus Manutius: 56
— member of the Florentine Academy: 58
— Zasius called 'alter Politianus': 86
— books containing handwritten notes by P.: 136
Pollux, ed. by Hemsterhuys: 153
Polybius, translated into Latin by Niccolò Perotti: 54
— Casaubon's notes on the text: 121
'polyhistory': 129, 138, 162, 187.2
polymathy *see* 'polyhistory'
Pompeii, excavations, and Winckelmann: 170
Pontano, Giovanni: 57
Pontanus, Jacobus: 77, 141
Pope, Alexander, translations of the *Iliad* and the *Odyssey*: 157 f.
Porphyry: 136
Porson, Richard: *159–61
— edition of Photius, *Bibliotheca*: 153
— review of Gibbon's *Decline and Fall of the Roman Empire*: 162
— 'lex Porsoni': 160, 179
— managed to avoid the burden of lecturing: 160, 181

Port-Royal, school of: 133
Postel, Guillaume de, and J. J. Scaliger: 114.4
Potter, John: 144
Prague, court of Charles IV: 58
Pralle, Ludwig, *Die Wiederentdeckung des Tacitus*: 31.1
prefaces to the *editiones principes* of ancient classics: 50.2
Presse Royale: 107 f., 111
Priapeia, and Boccaccio: 24
printing, in Italy: 50, 137 f.
— *editiones principes* of Greek and Latin classics: 50, 56; *see also editiones principes*
— Greek, in Venice: 50.2
— — in France: 107 f.
— Presse Royale: 107
— Aldine Press: 56
— offices of the Brethren of Common Life: 69
— in Basle: 83
— in Holland in the sixteenth and seventeenth centuries: 138
— Commelinus in Heidelberg: 141
— Wechel in Frankfurt: 141
Priscian: 52
Procopius, first edition of D. Höschel: 141
pronunciation, Greek: 53, 88 f. (Reuchlinian and Erasmian)
— — and Latin: 79 (Erasmus)
Propertius, ed. by Scaliger: 117
— ed. by Lachmann: 190
prose rythm, ancient, rediscovered by Bruni: 30
Psellus, ed. by Xylander: 141
Ptolemy, *Tetrabiblos*, ed. by J. Camerarius: 139
— *Almagest*, first Greek ed. by Camerarius and Grynaeus: 139
— translated by Regiomontanus: 37.8
Publilius Syrus, ed. by Bentley: 153
Pythagoras: 152

'Querelle des anciens et modernes': 100, 134, 150, 152, 158
Quintilian, the first complete manuscript discovered by Poggio: 32

Rabelais, F., and 'Demogorgon': 21
— in audience of *lecteurs royaux* for Greek: 102
— reader of Amyot's French Plutarch: 113
Ramboldi da Imola, Benvenuto, commentary on the *Commedia* of Dante: 23
Ramée, Pierre de la, and J. J. Scaliger: 114
ratio: 95, 154
rationalism: 41 (Valla), 74 (Erasmus), 145.2
realism: 190

reason, human, limits of: 75 (Erasmus)
— , and religion: 146 f. (Bentley and Boyle);
see also *ratio* and rationalism
'recensio': 190
Regiomontanus, and Bessarion: 37, 139.4
Reiske, Johann Jakob: *172
— and Hieronymus Wolf: 139 f.
— *Oratores Attici*: 140, 172
religio and *grammatica*: 116 (Scaliger)
religion, Greek, study of: 179 f. (Welcker);
see also mythology
religion and humanism: 183.1; see also
philosophia Christi
'Renaissance', origin of the term: 18
Reuchlin, Johannes: *86–90
— and the mss. of Erasmus used for his
edition of the NT: 76 f.
— and Budé: 101
Rhenanus, Beatus: *83 ff.
— the critic needs 'iudicium': 84, 137
rhetoric, Greek and Latin: 172
Rienzo, Cola di, and the origin of the Renaissance: 18
— at the court of the emperor Charles IV:
59
Rinucci, paraphrase of Aristophanes' *Plutus*
400–626: 29
— taught Poggio some Greek: 34
Ritschl, Friedrich, and J. Bernays: 119.7
Robortello, Francesco: *136 f.
— edition of Callimachus' *Hymns*: 46
— edition of Aeschylus: 111, 136
Roche-Pozay, family, and J. J. Scaliger:
115 f.
Roman:
— art: 170
— chronology and antiquities: 137
— civilization: 187
— culture: 170
— history, study of Niebuhr: 183 f.
— law, study of: 3, 43.5, 45 (Politian), 86
(Zasius), 101, 117 (Cuiacius), 184 (Savigny); see also *Digests* and *Pandects*
— literature, history of: 24 (Boccaccio,
Sicco Polentonus)
romantic movement, and Böckh: 181
romanticism, disliked by G. Hermann: 179
— influence on Niebuhr and K. O. Müller:
187
Rome, ancient, glory of: 6
— — antiquities, rediscovered by P. Leto:
51
— early, Niebuhr's understanding of its
history: 184
— modern, and Winckelmann: 168
— — and Niebuhr: 184
— — and Humboldt: 184 f.

— — Vatican Library: 49.5; see also
Academy, Roman
Ronsard, Pierre de, and the Pléiade: 102
— and Homer and Virgil: 103
— and Dorat: 105
Rubianus, Crotus: 90
Rüegg, Walter: 20
Ruhnken, David: 163

Sabbadini, Remigio: 19, 32
Saint-Germain-des-Prés, abbey: 131
Saint-Maur, Benedictine congregation of:
130 f.
Salel, H.: 134
Salmasius, Claudius: *122 f.
— and Scaliger: 119
— devoted attention to the study of Greek
dialects: 188
— *De lingua Hellenistica commentarius* and the
anonymous writing *Funus linguae Hellen-
isticae*: 123, 189.4
Salutati, Coluccio: *25–27
— interest in Plutarch: 29
— and Jean de Montreuil: 61.1
— knew the commentary of Donatus on
Terence: 155
Samxon, Jehan: 134
Sankt Gallen, and Poggio: 31 f.
— a palimpsest, identified by Niebuhr: 184
Sannazaro, Jacopo: 57
Sanskrit: 185
Savigny, Friedrich Karl von: 184
Savile, Henry: 144
Scaliger, Joseph Justus: *113–119
— and the authenticity of the works of
Dionysius Areopagita: 40
— on the importance of the knowledge of
Greek: 73
— four ages of Greek poetry: 119
— idea of an all-embracing scholarship of the
ancient world: 118, 137, 176, 183
— and Gruter: 116, 138
— and Hoeschel: 141
— and Winckelmann: 119, 168, 170
— and F. A. Wolf: 118, 176
— as man of religious personality never discussed the relation of Christianity to the
ancient world as a problem: 144
— and Roman history: 184
— discussions about 'lingua Hellenistica'
in the circle of his pupils: 189
Scaliger, Julius Caesar: 113 f.
Scheller, Meinrad: 85.4, 185.5
Schiller, Friedrich von: 173
Schlegel, Friedrich: *186
— humanistic approach fruitful in his
writings: 171

Schlegel, *contd.*
— enthusiastic review of Wolf's *Prolegomena*: 175
Schleicher, August: 185.5
Schleiermacher, Friedrich: *186
— his hermeneutic: 93
— Böckh his pupil: 181
— complete translation of Plato: 186
Schlettstadt: 83, 87
Schlick, Caspar: 60.1
scholarship, unity with poetry, *see* poetry
— and philosophy (Politian): 58
— and science: 114 (French Renaissance), 144 f. and 146 f. (Bentley, Boyle, and Newton)
— bulwark against ignorance: 79
— true: 82
— to be combined with understanding of art: 179 f.
— classical, *see also* 'Altertumswissenschaft'
— — and Lutheran exegesis of the Bible: 93
— — Böck's *Enzyklopädie und Methodologie der philologischen Wissenschaften*: 181 f.
— — two rival schools through the nineteenth into the twentieth century one concerned with words, the other with things: 182
— — history: 110 (H.Étienne, *De criticis veteribus Graecis et Latinis*)
— Hellenistic: 76
'Scholasticism': 17; *see also* philosophy, Scholastic
— condemned by Colet: 72
— Protestant: 82
Schopenhauer, Arthur: 34
Schütz, Chr. G.: 180.4
Schulpforta: 171 f.
Schwartz, Eduard: 134.1
science and scholarship: 114 (French Renaissance), 144 f. and 146 f. (Bentley, Boyle, and Newton); *cf. also* 167
Scipio Africanus, in Petrarch's *Africa* and *De viris illustribus*: 6 f.
script, Petrarch's feeling for beauty of: 13
— humanistic: 13
— 'Gothic', 'Gothico-Antiqua': 13
— Poggio and the 'littera antiqua': 32
— Ciriaco di Ancona and Felice Feliciano: 51.3
— Pontano's handwriting: 57.1
— Beneventan: 139
— *see also* minuscule, Carolingian
scriptoria, Greek, in Venice and in Crete: 56
Scriverius, Petrus: 115
Seehausen: 168
Selden, John, edition of the *Marmor Parium*: 119

— *Mare clausum*: 127
— and his generation: 144
Sellyng, William: 66
'Seminar', philological, created by Wolf at Halle: 176
Seneca, the contents of his plays described by A. Mussato: 4
— much in Petrarch's letters modelled on him: 10
— the authenticity of his correspondence with St. Paul: 40 (Valla), 75 (Erasmus)
— and Politian: 43
— Calvin's commentary on *De clementia*: 100
— edition of J. Lipsius: 126
Senfl, Ludwig, *Varia carminum genera*: 63.5
Sepulveda, Juan Ginéz: *94 f.
— about 'ratio': 95, 154.2
sermo, preferred to *verbum* by Erasmus: 74.4
Servius, commentary on Virgil: 3, 9, 14, 54.4
Seyssel, Claude de: 99, 113.2
Sforza, family, library: 49
Shakespeare, William, and Plutarch: 113
Shelley, P. B.: 21.7
Sicily, history, treated by Bentley: 152
sigla, used for manuscripts by Politian: 44
Sigonio, Carlo: 136 f.
Silius Italicus, *Punica*, rediscovered by Poggio: 7.2, 33
— — edition of H. Grotius: 127
Simler, Georg: 88
Simon, Richard: 130
Sixtus IV, pope: 52
'Society of Dilettanti': 161 f.
Sodalitas Φιλελλήνων in Venice: 56
Solinus, C. Iulius: 122
Sophocles, problem of the manuscript tradition: 111.8, 161
— Florentine manuscripts collated by Vettori: 136
— Cod. Laur. XXXII 9: 48, 111, 136
— edition of Turnebus: 111
— edition of W. Canter: 125
— Dorat's lecture on S.: 104
— Bentley on the spurious fr. 1126 P.: 150
Sorbonne: 71.5, 140
Spain, and Italian humanism: 61
— Greek studies at Alcalà: 65
— 'Erasmianism' in Spanish cultural life: 94 ff.
Spalding school: 145
Spenser, Edmund: 144
— *Faery Queene*: 21
Speyer: 85
Spinoza, Baruch: 128.6
Spon, Jacques: 133
Stanley, Thomas: 144
— notes to Callimachus: 144, 152

— criticism of Aeschylus: 161
Statius, *Silvae*, discovered by Poggio: 33
— — and Politian: 43
Steinthal, H. 185.2
Stendal: 167
Stendhal (Henri Beyle): 167.5
Stephanus, *see* Étienne
Stephenses, *see* Étienne
Steyn, Augustinian monastery: 71
Stillingfleet, Dr.: 145, 149
Stoicism, Christian, proclaimed by J. Lipsius: 126
Strabo, *Geographica*, commentary of Casaubon: 121
Strassburg: 83, 185
studia humanitatis see humanitas
'Studio fiorentino': 56
Stunica, Jacobus *see* Zuñiga
Sturm, Jacob: 83
Suetonius, *Vita* of Terence: 7
— codex of S. used by Petrarch: 13
— edited by Erasmus: 77
— definition of φιλόλογος: 101
— commentary of Casaubon: 121
— *Caesares*, Bentley's contributions to the text: 155
Suidas, ed. by L. Küster: 150, 153, 162
Swift, Jonathan: 144, 150.6, 151
Swiss, humanism: 85 f.
Synesius, ed. by D. Petau (1612): 132
Sylburg, Friedrich: 141
Synkellos, Georgios: 118

Tabula Peutingeriana: 64, 141
Tacitus, the codex Casinensis and Boccaccio: 23
— rediscovery of the minor works and the codex Fuldensis: 31
— the codex Mediceus I, found in Corvey (1508): 85.1
— *Germania*, ed. by Celtis: 64
— — ed. by B. Rhenanus: 84
— 'Tacitean movement' across Europe and the edition of Lipsius: 126
teaching in the Italian Renaissance: 52 ff.
teleology: 147 (Bentley)
Temple, Sir William: 150 f.
Terence (Terentius Afer), confusion with Terentius Culleo corrected by Petrarch: 7
— ed. by Erasmus: 77
— ed. by Bentley: 153, 155
— commentary of Donatus used by Bentley: 155
'*Textgeschichte*': 130 (Bible), 174 (Homer)
Theocritus, edition of Aldus: 56.4
— edition of Ahrens: 188
— Politian's lectures on Th.: 44

— and Dorat: 104
Theodontius: 22.2, 4
Theognis, manuscripts, and J. Lascaris: 48.8
— with other Greek gnomic poets edited by Turnebus: 111
Theognost, *Orthography*, discovered by Bentley: 149
theology, of Erasmus: 71.3
Theophrastus, *Characteres*, commentary of Casaubon: 121
Thesaurus Linguae Latinae: 107.4
Thirty Years' War: 142
Thomas a Kempis, educated in a Deventer school: 69
Thomas Aquinas, and Erasmus: 74
Thorwaldsen, Bertel: 179
Thou, Jacques Auguste de: 117, 119, 120, 132, 168
Thucydides, VI 2.4: 9
— VIII 328: 9
— and Valla: 38 f.
Tibullus, ed. by Scaliger: 117
— ed. by Lachmann: 190
Tillemont, Lenain de, Sébastien, *see* Lenain de Tillemont, Sébastien
Tissard, François: 104
Toffanin, G., against the 'paganism' of the Italian Renaissance: 19
Tortelli, Giovanni, on *Orthography*: 55
— and Poggio: 55.3
Toup, Jonathan; work on Suidas: 153, 162
Tournefort, Joseph Pitton de: 155.7
Toussain, Jacques: 102 f., 111
tragedies, Greek, lost, reconstructed by Welcker: 180
translation:
— Bruni as translator of Greek literature and his *De interpretatione recta*: 28 f.
— St. Jerome, *Ep.* 106 (his principles of translating): 29.7
— translations of classical literature in England: 143 f.
Travis, George: 160
Triclinius, recension of Aeschylus: 111
Trieste: 168
Tritonius, Petrus: 63 f.
truth, poetry must be based on (Hesiod-Ennius-Petrarch): 7
— *veritas* (Valla): 35 f., 37
Tübingen, university, Reuchlin's lectures on Greek and Hebrew: 87
Turnebus, Adrianus: *111 f.
— *lecteur royal* at the Collège Royal: 103
— *Adversaria*: 111
— and J. J. Scaliger: 115
— and W. Canter: 125
Tyrwhitt, Thomas: 161 f.

Ullman, Berthold Louis: 19
universities of the Italian Renaissance: 56
Upper Rhine, humanistic circle: 82 ff.
Urbino, ducal library: 50
Urceus, Codrus: 55

Valckenaer, L. C.: 163
Valerius Flaccus, *Argonautica*, manuscript
    discovered by Poggio: 32
Valerius Maximus, commentary of Dionigi
    de' Roberti: 54
— edition of Lipsius: 126
Valla, Lorenzo: *35–41
— and Poggio: 34, 35, 90
— and the term 'Gothic': 35, 84
— critical spirit: 38, 75 f.
— rationalism: 145.2
— *Emendationes Livianae*: 36, 184
— and Erasmus: 75 f.
— influence on Seyssel: 99.4
— interested in ancient fables: 91
— 'loci' in his *Dialecticae disputationes*: 92
Varro, *De lingua Latina*, Cod. Laur. 50.10:
    24
— — ed. by A. Agustín: 95
— — copies brought to France by J. de
    Montreuil: 65
— — J. J. Scaliger's *Coniectanea*: 115
Velleius Paterculus, first edition of B.
    Rhenanus: 85
— edition of J. Lipsius: 126
Venice, Greek printing: 50.2; *see also*
    Manutius, Aldus
— Greek scriptoria: 56
— 'Sodalitas Φιλελλήνων': 56
— and Erasmus: 73
— and P. Leto's teaching: 51
— seat of humanism: 88
Verbum and *sermo* (Erasmus): 74.4
Verona, Capitular Library: 184
Vettori, Piero: *135 f.
— edition of Aeschylus *Prometheus*: 104, 111
— his books and manuscripts in the Bavarian
    State Library, Munich: 139
vetustas: 15
Vibius Sequester: 22
Victorius, Petrus, *see* Vettori, Piero
Vienna: 163
Villoison, J.–B. d'Ansse de: 174
Virgil, commentary of Servius: 3, 9, 14, 54.4
— and Petrarch: 6, 9
— and Boccaccio: 20
— Pomponio Leto's 'notes on V.': 55
— Renaissance commentaries: 55.6
— 'la curieuse diligence de Virgile' (Ron-
    sard): 103
— and Bentley: 146, 158

— Homer, not V. the source of inspiration in
    the age of Neohellenism: 167
— and Homer: 174
— *see also Appendix Virgiliana*
Visconti, family, library: 49
Visconti, Gran Galeazzo: 26
'vita activa' (Salutati): 25
Vittorino da Feltre: 53.3
Vivès, Juan Luis: *96
— his Latin books recommended by pope
    John XXIII: 77
Voigt, Georg, *Die Wiederbelebung des classischen
    Alterthums oder das erste Jahrhundert des
    Humanismus*: 17 f.
Vondel, Joost van den: 128
Voorbrock, Jacob, *see* Perizonius, Jacobus
Vossius, Gerard John: *137
— *De philologia*: 129
Vossius, Isaac: 129
Vulcanius, Bonaventura: 153

Wagner, Richard, and Droysen: 189
Walser, Ernst, Renaissance studies: 18
— biography of Poggio: 18 f., 31.2
wars of religion, French: 113 f.
Wechel, printer and publisher in Frankfurt:
    141
Welcker, Friedrich Gottlieb: *179 f.
— Niebuhr's depreciatory remarks on him:
    184.4
— and Droysen: 188
Welser, Marcus, editor of the *Tabula Peutin-
    geriana*: 64, 141
Wetstein, Johann Jacob, and Bentley: 156
Wheler, G.: 133.2
Wilamowitz-Moellendorff, Ulrich von, on
    Valla's *Declamatio*: 39
— running commentaries and monographs:
    79
— and Dorat: 103.3
— and R. Simon: 130.4
— and F. A. Wolf: 177
— and G. Hermann: 179
— Latin studies: 178.2
— edition of the *Bucolic poets* and his judge-
    ment of Ahrens's edition: 188
will, free, problem of (Luther-Erasmus):
    80
Wimpfeling, Jacob: 83
Winckelmann, Johann Joachim: *167–72
— and Scaliger: 119, 170, 176
— biography written by F. A. Wolf: 173 f.
— influenced Böckh: 181
— ideas and writings decisive for the future
    of classical scholarship: 183
— Goethe, *Winckelmann und sein Jahrhundert*:
    169, 185

Wittenberg, university, and Melanchthon: 91, 93 f.
Wölfflin, Eduard: 107.4
Wolf, Friedrich August: *173–7
— conception of 'Altertumswissenschaft' anticipated by Scaliger: 118 and Robortello: 137
— and R. Simon: 130.4
— sketch of a life of Bentley: 148.8, 173
— not able to grasp the significance of Bentley's discovery of the digamma: 157
— and Böckh: 180
— and I. Bekker: 180
— and K. O. Müller: 186
Wolf, Hieronymus: *139 f.
— pupil of Melanchthon: 94
— and Georgios Synkellos: 118.3
— and R. Ascham: 143
— and Reiske: 140, 172
Wolff, Emil: 146
Wood, Robert: 161, 175
Worcester: 145

Worms: 85
Wotton, William: 147, 151
Wren, Christopher: 147
Wyttenbach, Daniel: 163

Xenophon, translation into Latin by Bruni: 29
— translation of smaller excerpts by Valla: 38
— translated by Erasmus: 77
Ximenes de Cisneros, Francisco: 65, 95.2
Xylander see Holtzmann

Zancle: 152
Zasius, Ulrich: 86
Zenodotus: 54
Zoëga, Johann Georg: 179
Zonaras, Ioannes, first edition of H. Wolf: 140
Zuñiga, Jacobus Lopis: 94
Zwingli: 73

INDEX OF GREEK WORDS

ἀρχαιολογία: 133.3
γραμματική, see grammar
γραμματικός: 52, 58.5
ἑλληνίζειν, ἑλληνιστής, ἑλληνισμός: 189
ἴαμβος: 90
λόγος, 'sermo', preferred to 'verbum' by Erasmus: 74.4
μαντική: 154
μίμησις, problem of, and Winckelmann: 169

παιδεία = 'humanitas': 15
πολυΐστωρ: 129.4
πρώκιον: 188.1
συγγράμματα: 79
ὑποθέσεις: 4
ὑπομνήματα: 79
φιλανθρωπία = 'humanitas': 15, 17
φιλόλογος: 101
χρονογραφία: 149

INDEX OF PASSAGES DISCUSSED

Bible, OT, *Eccles.* 18:6: 16 n. 2
— NT, St. John 21: 22 ff. (treatment of Bessarion): 37
— — I St. John 5: 7 and 8 (Comma Ioanneum): 148, 160
Callimachus, *hy.* V 136: 54 f.
— fr. 21.3 (Bentley's treatment): 153 n. 4.

Erasmus, *Ep.* 46.41: 65 n. 2
Livy, XXIII 30.3 and 34.17: 37
[Theocr.] VIII 91: 188 n.1
Schol. [Theocr.] IV 16a: 188 n. 1
Virgil, *Aen.* I 29 ff. and II 254: 9
— — X 851: 6 n. 3

OTHER TITLES IN THIS HARDBACK REPRINT PROGRAMME FROM SANDPIPER BOOKS LTD (LONDON) AND POWELLS BOOKS (CHICAGO)

ISBN 0–19–	Author	Title
8143567	ALFÖLDI A.	The Conversion of Constantine and Pagan Rome
6286409	ANDERSON George K.	The Literature of the Anglo-Saxons
8228813	BARTLETT & MacKAY	Medieval Frontier Societies
8111010	BETHURUM Dorothy	Homilies of Wulfstan
8142765	BOLLING G. M.	External Evidence for Interpolation in Homer
9240132	BOYLAN Patrick	Thoth, the Hermes of Egypt
8114222	BROOKS Kenneth R.	Andreas and the Fates of the Apostles
8203543	BULL Marcus	Knightly Piety & Lay Response to the First Crusade
8216785	BUTLER Alfred J.	Arab Conquest of Egypt
8148046	CAMERON Alan	Circus Factions
8148054	CAMERON Alan	Porphyrius the Charioteer
8148348	CAMPBELL J.B.	The Emperor and the Roman Army 31 BC to 235 AD
826643X	CHADWICK Henry	Priscillian of Avila
826447X	CHADWICK Henry	Boethius
8219393	COWDREY H.E.J.	The Age of Abbot Desiderius
8148992	DAVIES M.	Sophocles: Trachiniae
825301X	DOWNER L.	Leges Henrici Primi
814346X	DRONKE Peter	Medieval Latin and the Rise of European Love-Lyric
8142749	DUNBABIN T.J.	The Western Greeks
8154372	FAULKNER R.O.	The Ancient Egyptian Pyramid Texts
8221541	FLANAGAN Marie Therese	Irish Society, Anglo-Norman Settlers, Angevin Kingship
8143109	FRAENKEL Edward	Horace
8201540	GOLDBERG P.J.P.	Women, Work and Life Cycle in a Medieval Economy
8140215	GOTTSCHALK H.B.	Heraclides of Pontus
8266162	HANSON R.P.C.	Saint Patrick
8224354	HARRISS G.L.	King, Parliament and Public Finance in Medieval England to 1369
8581114	HEATH Sir Thomas	Aristarchus of Samos
8140444	HOLLIS A.S.	Callimachus: Hecale
8212968	HOLLISTER C. Warren	Anglo-Saxon Military Institutions
8223129	HURNARD Naomi	The King's Pardon for Homicide – before AD 1307
8140401	HUTCHINSON G.O.	Hellenistic Poetry
9240140	JOACHIM H.H.	Aristotle: On Coming-to-be and Passing-away
9240094	JONES A.H.M	Cities of the Eastern Roman Provinces
8142560	JONES A.H.M.	The Greek City
8218354	JONES Michael	Ducal Brittany 1364–1399
8271484	KNOX & PELCZYNSKI	Hegel's Political Writings
8225253	LE PATOUREL John	The Norman Empire
8212720	LENNARD Reginald	Rural England 1086–1135
8212321	LEVISON W.	England and the Continent in the 8th century
8148224	LIEBESCHUETZ J.H.W.G.	Continuity and Change in Roman Religion
8141378	LOBEL Edgar & PAGE Sir Denys	Poetarum Lesbiorum Fragmenta
9240159	LOEW E.A.	The Beneventan Script
8241445	LUKASIEWICZ, Jan	Aristotle's Syllogistic
8152442	MAAS P. & TRYPANIS C.A .	Sancti Romani Melodi Cantica
8142684	MARSDEN E.W.	Greek and Roman Artillery—Historical
8142692	MARSDEN E.W.	Greek and Roman Artillery—Technical
8148178	MATTHEWS John	Western Aristocracies and Imperial Court AD 364–425
8223447	McFARLANE K.B.	Lancastrian Kings and Lollard Knights
8226578	McFARLANE K.B.	The Nobility of Later Medieval England
8148100	MEIGGS Russell	Roman Ostia
8148402	MEIGGS Russell	Trees and Timber in the Ancient Mediterranean World
8142641	MILLER J. Innes	The Spice Trade of the Roman Empire
8147813	MOORHEAD John	Theoderic in Italy
8264259	MOORMAN John	A History of the Franciscan Order
8116020	OWEN A.L.	The Famous Druids
8131445	PALMER, L.R.	The Interpretation of Mycenaean Greek Texts
8143427	PFEIFFER R.	History of Classical Scholarship (vol 1)
8143648	PFEIFFER Rudolf	History of Classical Scholarship 1300–1850
8111649	PHEIFER J.D.	Old English Glosses in the Epinal-Erfurt Glossary
8142277	PICKARD–CAMBRIDGE A.W.	Dithyramb Tragedy and Comedy
8269765	PLATER & WHITE	Grammar of the Vulgate
8213891	PLUMMER Charles	Lives of Irish Saints (2 vols)
820695X	POWICKE Michael	Military Obligation in Medieval England
8269684	POWICKE Sir Maurice	Stephen Langton
821460X	POWICKE Sir Maurice	The Christian Life in the Middle Ages
8225369	PRAWER Joshua	Crusader Institutions

8225571	PRAWER Joshua	The History of The Jews in the Latin Kingdom of Jerusalem
8143249	RABY F.J.E.	A History of Christian Latin Poetry
8143257	RABY F.J.E.	A History of Secular Latin Poetry in the Middle Ages (2 vols)
8214316	RASHDALL & POWICKE	The Universities of Europe in the Middle Ages (3 vols)
8154488	REYMOND E.A.E & BARNS J.W.B.	Four Martyrdoms from the Pierpont Morgan Coptic Codices
8148380	RICKMAN Geoffrey	The Corn Supply of Ancient Rome
8141076	ROSS Sir David	Aristotle: Metaphysics (2 vols)
8141092	ROSS Sir David	Aristotle: Physics
8142307	ROSTOVTZEFF M.	Social and Economic History of the Hellenistic World, 3 vols.
8142315	ROSTOVTZEFF M.	Social and Economic History of the Roman Empire, 2 vols.
8264178	RUNCIMAN Sir Steven	The Eastern Schism
814833X	SALMON J.B.	Wealthy Corinth
8171587	SALZMAN L.F.	Building in England Down to 1540
8218362	SAYERS Jane E.	Papal Judges Delegate in the Province of Canterbury 1198–1254
8221657	SCHEIN Sylvia	Fideles Crucis
8148135	SHERWIN WHITE A.N.	The Roman Citizenship
9240167	SINGER Charles	Galen: On Anatomical Procedures
8113927	SISAM, Kenneth	Studies in the History of Old English Literature
8642040	SOUTER Alexander	A Glossary of Later Latin to 600 AD
8222254	SOUTHERN R.W.	Eadmer: Life of St. Anselm
8251408	SQUIBB G.	The High Court of Chivalry
8212011	STEVENSON & WHITELOCK	Asser's Life of King Alfred
8212011	SWEET Henry	A Second Anglo-Saxon Reader—Archaic and Dialectical
8148259	SYME Sir Ronald	History in Ovid
8143273	SYME Sir Ronald	Tacitus (2 vols)
8200951	THOMPSON Sally	Women Religious
8201745	WALKER Simon	The Lancastrian Affinity 1361–1399
8161115	WELLESZ Egon	A History of Byzantine Music and Hymnography
8140185	WEST M.L.	Greek Metre
8141696	WEST M.L.	Hesiod: Theogony
8148542	WEST M.L.	The Orphic Poems
8140053	WEST M.L.	Hesiod: Works & Days
8152663	WEST M.L.	Iambi et Elegi Graeci
822799X	WHITBY M. & M.	The History of Theophylact Simocatta
8206186	WILLIAMSON, E.W.	Letters of Osbert of Clare
8114877	WOOLF Rosemary	The English Religious Lyric in the Middle Ages
8119224	WRIGHT Joseph	Grammar of the Gothic Language